UNITEXT for Physics

UNITEXT for Physics series, formerly UNITEXT Collana di Fisica e Astronomia, publishes textbooks and monographs in Physics and Astronomy, mainly in English language, characterized of a didactic style and comprehensiveness. The books published in UNITEXT for Physics series are addressed to graduate and advanced graduate students, but also to scientists and researchers as important resources for their education, knowledge and teaching.

More information about this series at http://www.springer.com/series/13351

Sergio Petrera

Problems and Solutions in Nuclear and Particle Physics

 Springer

Sergio Petrera
Gran Sasso Science Institute
L'Aquila, Italy

ISSN 2198-7882　　　　　　　ISSN 2198-7890　(electronic)
UNITEXT for Physics
ISBN 978-3-030-19775-9　　　ISBN 978-3-030-19773-5　(eBook)
https://doi.org/10.1007/978-3-030-19773-5

This Springer imprint is published by the registered company Springer Nature Switzerland AG
The registered company address is: Gewerbestrasse 11, 6330 Cham, Switzerland

Preface

"The reader who has read the book but cannot do the exercises has learned nothing." I used this quote by J. J. Sakurai [1] at each and every first lesson of my courses on nuclear and particle physics. In my message to the students, the "book" in the quote represented the course they were attending. In this way, I wanted to place special emphasis on the importance of exercises in such introductory courses. There are several good textbooks that I used as the basis of my courses and that I have proposed to my students, yet the exercises proposed therein are only partly solved or simply sketched. For this reason, I used to tell my students that even if they missed some lessons, they should follow closely the lessons dedicated to working out problems.

This book contains a sample of about 140 solved problems on nuclear and particle physics. These problems have been used in partial and final examinations of courses I have given for about twenty years, mostly to undergraduates in the University of L'Aquila. In these lecture notes, the solutions are explained in detail and different approaches are proposed and sometimes compared. Another feature of the exercises originates from the decision to consider only realistic cases, to have solutions as close as possible to what is available from actual measurements. Whenever possible, some problems are based on well-known experiments to show that even with their basic knowledge students can understand the main outcomes of these researches.

The exercises are grouped by different subjects. This grouping criterion is not (and cannot be) rigorous because a generic exercise needs inputs from different topics. Therefore, the exercises included in each chapter refer to it only because this is the prevalent subject. The levels of the exercises and the required skills to solve them can be very different. Most of the exercises do not require too much mathematics. Yet, some of the exercises are more difficult and complex and serve as prototypes for a class of problems, so that others of the same class can be solved promptly.

Apart from the use of this book as a supplement to textbooks on nuclear and particle physics for undergraduate classes, it can provide a valid aid to graduate students preparing for selection examinations.

L'Aquila, Italy   Sergio Petrera
April 2019

Reference

1. Foreword by J. S. Bell. In: Sakurai, J.J.: Modern Quantum Mechanics. Addison-Wesley (1994)

The original version of the book was revised: References have been added to the preface of the book. The correction to the book is available at https://doi.org/10.1007/978-3-030-19773-5_4

Acknowledgements

Several young collaborators helped me to prepare and check the problems along the courses I have given. I wish to thank all of them and, in particular, Eugenio Scapparone and Francesco Salamida. I also would like to thank Celina Paul for her help with the English language.

Contents

Notes

Data

Each problem can be taken as stand-alone. This means that all input data are provided in the text: For example, the relevant particle masses are usually given in the text. The reader may notice that their accuracies can change on a case-by-case basis. This feature is a consequence of the origin of the text, since these problems were used for examinations and I preferred to give all the needed input data at the accuracy required for each specific case. On the other hand, it also allows the reader to pick up problems randomly without requiring a sequential reading.

The problems are mainly numerical and require values of physical constants, especially for conversion purposes. Whenever these values are not reported in the text, the reader can refer to the PDG Review of Particle Physics [2] which provides an up-to-date collection of constants, units, atomic, and nuclear properties. This review is much more than a simple collection and can be considered as a "must" for dealing with any nuclear and particle physics case.

Nuclear physics data are available from several sources. Some examples are the National Nuclear Data Center (NNDC) at Brookhaven National Laboratory [3] and the IAEA Nuclear Data Section [4].

Units

We use the International System of Units (SI), except for energy, mass, and momentum which are specified in terms of eV. This mixed system can be easily handled and the system-specific electromagnetic constants disappear promptly, using the SI definition of the fine structure constant α and the value of $\hbar c$ in mixed units.

In nuclear physics, kinematical expressions are mostly non-relativistic. In particle physics, the relativistic treatment is instead mandatory. As adopted in many books, in all kinematical expressions c is omitted (i.e., $c = 1$), making them simpler to be handled. Once the energy scale of the problem is set, e.g., GeV, the right units are easily restored with the rule that momenta, energies, and masses are finally given in GeV/c, GeV, and GeV/c^2 respectively. For all the other quantities (e.g., velocity, time, distance, etc.), the light velocity c is kept in the equations.

Other References

There are several excellent books that deal with either nuclear or particle physics. Less frequently does one see textbooks presenting these two areas of physics in a unified manner, especially at the undergraduate level. The books Nuclear and Particle Physics by W. S. C. Williams [5], Particles and Nuclei by B. Povh et al. [6], Nuclear and Particle Physics by B. R. Martin [7], and Introduction to Nuclear and Particle Physics by A. Das and T. Ferbel [8] provide the kind of combined exposition more appropriate to the level of the problems proposed here. Finally, a very useful collection of solved problems, including also different topics, is Problems and Solutions on Atomic, Nuclear and Particle Physics by Yung-Kuo Lim [9].

References

1. Tanabashi, M., et al.: Particle data group. Phys. Rev. D **98**, 030001 (2018). http://pdg.lbl.gov/
2. National Nuclear Data Center (NNDC) at Brookhaven National Laboratory (https://www.nndc.bnl.gov/)
3. IAEA Nuclear Data Section (https://www-nds.iaea.org/relnsd/vcharthtml/VChartHTML.html)
4. Williams, W.S.C.: Nuclear and Particle Physics. Clarendon Press, Oxford (1991)
5. Povh, B., Rith, K., Scholz, C., Zetsche, F.: Particles and Nuclei. Springer (1993)
6. Martin, B.R.: Nuclear and Particle Physics. Wiley (2006)
7. Das, A., Ferbel, T.: Introduction to Nuclear and Particle Physics. World Scientific (2003)
8. Lim, Y.-K.: Problems and Solutions on Atomic, Nuclear and Particle Physics. World Scientific (2000)

Chapter 1
Nuclear Physics

Abstract This chapter is dedicated to Nuclear Physics. After a few very simple problems, it addresses nuclear scattering, the binding energy of nuclei, nuclear decays and nuclear models. Most of the formulas used here are based on the book by Williams (Nuclear and Particle Physics. Clarendon Press, Oxford, 1991) [1], but sometimes the expressions from other books are preferred, when they lead to simpler solutions. In fact, some parametric formulas, e.g. the nuclear radius dependence on the mass number, the *semi-empirical mass formula*, etc., can differ from text to text and the associated parameters change accordingly.

1.1 Initial Problems

Exercise 1.1.1

Estimate the nuclear density in g/cm^3.

Exercise 1.1.2

Using only classical electromagnetism, give an estimate of the Coulomb term (a_C) in the *semi-empirical mass formula (SEMF)*.

Exercise 1.1.3

A neutron star is an astrophysical object with a density similar to the one of a nucleus. Knowing that its typical mass is of the order of one solar mass ($M_\odot = 2 \times 10^{30}$ kg), calculate its radius.

Exercise 1.1.4

A deuteron gas (a *deuteron* is a nucleus of deuterium, $^2_1$H) is heated at temperature T. For which temperature nuclear processes occur? Which is the interaction involved? [$k_B = 8.6 \times 10^{-5}$ eV/K].
Hint: *nuclear interaction is possible if the distance between deuterons is of the order of 1 fm.*

© Springer Nature Switzerland AG 2019   1
S. Petrera, *Problems and Solutions in Nuclear and Particle Physics*,
UNITEXT for Physics, https://doi.org/10.1007/978-3-030-19773-5_1

Exercise 1.1.5

A gaseous tritium ($^3_1$H) target is bombarded with a mono-energetic deuteron ($^2_1$H) beam. The tritium nuclei can be assumed at rest. In the collisions α particles and neutrons are produced, through the reaction

$$^2_1\text{H} +^3_1\text{H} \rightarrow ^4_2\text{He} + n$$

What is the neutron rate (neutrons per sec) in a detector at $\theta = 30°$ having section $S = 20$ cm^2 and distance $R = 3$ m from the target?

The target thickness is $L\rho = 0.2$ mg/cm^2, the differential cross section is $d\sigma/d\Omega(30°) = 13$ mb/sr. The beam intensity is $I = 2\,\mu$A.

1.2   Nuclear Scattering

Exercise 1.2.1

(1) A 5 μA electron beam with momentum 700 MeV/c is incident upon a ^{40}Ca target thick 0.12 g/cm^2. A detector having section 20 cm^2 far 1 m from the target is positioned at 40° with respect to the beam direction to measure the scattered electrons. Assuming that the charge distribution of the nucleus is uniform in a sphere of radius $(1.18\,A^{1/3} - 0.48)$ fm, calculate the rate of the electrons hitting the detector.

(2) The detector is moved and positioned at 25°, where is the first local maximum of the differential cross section. Here the detector collects about 1400 counts per second. The detector consists in two gas counters in sequence, 1 mm thick each, filled with an Ar/CO$_2$ mixture. In this gas the electron energy loss is about 1.4 times the *ionization minimum*, the density is 1.8 mg/cm^3, the ionization potential is 15 eV. Assume that about 10% of the energy deposit is effectively converted into electron-ion pairs. Each detector provides a count even if a single electron reaches the anode, being the probability for the electron to reach the anode P $\simeq$ 30%. An event is recorded when the two counters provide signals in coincidence. Estimate the counting rate.

Exercise 1.2.2

Electrons with energy 180 MeV are elastically scattered by an ^{197}Au target. The angular distribution has a typical diffractive behaviour with several local maxima and minima. Assuming that the nucleus is a hard uniform sphere, evaluate the number of minima.

Exercise 1.2.3

In the Geiger and Marsden experiment the scattered α particles were counted observing the light flashes produced in a ZnS detector put in a movable ocular looking at the target. A human is able to count up to a maximum rate of few per second. Assuming that this rate is achieved when the ocular is positioned at 20°, what is the beam attenuation needed to extend the measurements down to 10°? Using the attenuated beam, what is the mean waiting time between two flashes at 20°?

Exercise 1.2.4

An electron beam with momentum 100 MeV/c and intensity $I_0 = 10$ μA hits a carbon target 1 g/cm² thick. A detector of section $S = 30$ cm² is positioned at 15° at a distance $R = 2$ m from the target. Calculate the rate of scattered electrons.

Exercise 1.2.5

500 MeV electrons are elastically scattered through an angle of 10° by Fe nuclei ($A = 56$). Calculate:

– the momentum transfer;
– the Mott cross section;
– the differential cross section for a uniform charge distribution.

Exercise 1.2.6

We aim to repeat the Geiger and Marsden experiment in a science lab. For this purpose the following items are available:

• an ^{241}Am source emitting α particles with 5.5 MeV kinetic energy;
• a thin gold foil as target ($A = 197$, $Z = 79$) having $\rho \Delta l = 0.1$ g/cm²;
• a detector with associated electronics to discriminate and count α particles and a computer to read out data. The detector has a sensitive surface of 10 cm² and can be positioned at different angles, keeping the same distance (1 m) from the target.

To achieve a good measurement of the cross section between 10° and 150° we require to count at least 10 α/s. What is the (minimum) intensity of α particles on target to achieve the required accuracy?

Exercise 1.2.7

Consider the reaction $p + {}^7_3\text{Li} \rightarrow {}^4_2\text{He} + {}^4_2\text{He}$. The binding energies of ${}^4_2\text{He}$ and ${}^7_3\text{Li}$ are 28.3 MeV and 39.3 MeV respectively.

– Establish if the reaction is either exothermic ($Q > 0$) or endothermic ($Q < 0$)
– Evaluate the spin-parity of the ${}^7_3\text{Li}$ nucleus.
– Assuming the ${}^7_3\text{Li}$ at rest, calculate the minimum proton energy for the reaction to occur.
– Knowing that the final angular momentum is null, calculate the initial angular momentum of the $(p, {}^7_3\text{Li})$ system. [The proton parity is $+$].

Exercise 1.2.8

Expanding the nuclear form factor in series of q^2 powers, one gets

$$F(q^2) = 1 - \frac{q^2}{6\hbar^2} \langle r^2 \rangle + \ldots \tag{1.1}$$

A measurement of the differential cross section of electrons, scattered elastically through 5°, gives 80 mb/sr for incident electrons with momentum 720 MeV/c upon carbon target.

1. Compare the measured cross section to the Mott formula.
2. Estimate the nuclear radius from (1.1).

Exercise 1.2.9

In his paper on Nature in 1932, Chadwick [2] motivates the discovery of the neutron arguing that the 'penetrating' neutral particle, obtained bombarding the nucleus of ^{9}Be with α particles, cannot be a gamma (as supposed earlier), that is

$$\alpha + {}^9\text{Be} \rightarrow {}^{13}\text{C} + \gamma, \tag{1.2}$$

but is instead a neutral particle having approximately the same mass of a proton, the 'neutron'

$$\alpha + {}^9\text{Be} \rightarrow {}^{12}\text{C} + n. \tag{1.3}$$

These particles are studied through their scattering against protons

$$\gamma(n) + p \rightarrow \gamma(n) + p. \tag{1.4}$$

Chadwick reports that the scattered protons have β not exceeding 0.1.
Show that:

(a) in the case of photons from reaction (1.2), the protons scattered in reaction (1.4) cannot have energies corresponding to the measured velocity, if their energies are of the order of 10 MeV, as expected from the α energy and the mass difference between initial and final nuclei.
(b) To have protons with the observed energies, photons in (1.2) should have an energy of $\approx$50 MeV.
(c) If instead the neutral particles are neutrons as in (1.3), the scattered protons are consistent with the measurements.

Exercise 1.2.10

To repeat the Geiger and Marsden experiment we use

- an ^{241}Am source emitting α particles with $T_\alpha = 5.64$ MeV;
- a 50 μm thick gold foil as target ($A = 197$, $Z = 79$, $\rho = 19.3$ g/cm^3);
- a detector of section 0.5 cm^2 and distance 10 cm, which is moved at seven different angles to count the scattered α particles.

After one hour of measurements at each angle, we collect the counts reported in the table below.

Calculate the intensity of α particles and its uncertainty.

θ	15°	25°	35°	45°	55°	65°	75°
counts per hour	4265	594	149	50	31	13	7

1.3 Nuclear Binding Energy

Exercise 1.3.1

Among the $A = 197$ isobars, the nucleus $^{197}_{79}$Au is stable. Which are the expected radioactive decay types for $^{197}_{78}$Pt and $^{197}_{80}$Hg to $^{197}_{79}$Au?

Exercise 1.3.2

Thermal neutrons (i.e. neutrons in thermal equilibrium with the medium) can induce the following fission reaction

$$^{235}_{92}U + n \longrightarrow \; ^{148}_{57}La + \; ^{87}_{35}Br + n$$

Assuming that the medium temperature is 300 K, estimate the energy released in the reaction.

Exercise 1.3.3

Deuterium ($^{2}_{1}$H) and tritium ($^{3}_{1}$H) nuclei have binding energies of 2.23 MeV and 8.48 MeV respectively. What is mean kinetic energy of the nuclei to bring them at a distance of 1.4 fm? What is the corresponding temperature?
   In this thermal condition the following reaction can occur

$$^{2}_{1}H + \; ^{2}_{1}H \longrightarrow \; ^{3}_{1}H + p$$

Calculate the energy release per reaction.

Exercise 1.3.4

The Sun is a copious source of neutrinos (*solar neutrinos*). The first observation of these neutrinos has been achieved in 1978 by R. Davis [3] in the Homestake mine (USA), using a large detector filled with C_2Cl_4. The reaction used for the detection is

$$\nu_e + \; ^{37}_{17}Cl \rightarrow \; ^{37}_{18}Ar + e^-.$$

Calculate the threshold energy of the reaction.
N.B. - Assume both nuclei in their ground state. The following numerical values are needed for the calculation $M_p - M_n = -1.293$ MeV/c^2, $m_e = 0.511$ MeV/c^2 and

the Coulomb and asymmetry coefficients appearing in the *SEMF*, $a_C = 0.697$ MeV, $a_A = 23.3$ MeV.

Exercise 1.3.5

Using the *semi-empirical mass formula* establish if the nucleus ${}^{64}_{29}$Cu can have β^- decay (into ${}^{64}_{30}$Zn) and/or β^+ decay (into ${}^{64}_{28}$Ni). Calculate also the maximum energies of the emitted $e^\pm$ [$M_p = 938.272$ MeV/c^2, $M_n = 939.565$ MeV/c^2, $m_e = 0.511$ MeV/c^2.]

Exercise 1.3.6

The most stable nucleus with $A = 101$ is ${}^{101}_{44}$Ru. Using this knowledge, the *semi-empirical mass formula* and the Coulomb coefficient in this formula, asymmetry [$M_n - m_e - M_p = 0.782$ MeV/c^2, $a_C = 0.697$ MeV].
Hint: The stability condition can be expressed as a minimum of the atomic mass.

Exercise 1.3.7

The maximum kinetic energy of the positrons from the β^+ decay of ${}^{35}_{18}$Ar is 4.95 MeV. Estimate the Coulomb term a_C as defined in the *semi-empirical mass formula*. Compare this value with the best-fit value of the *SEMF*.
[$M_p - M_n = -1.293$ MeV/c^2, $m_e = 0.511$ MeV/c^2]

Exercise 1.3.8

Confirm with the *SEMF* that ${}^{100}_{43}$Tc can transmute to both ${}^{100}_{44}$Ru and ${}^{100}_{42}$Mo. Which are the corresponding transitions?
[$M_p = 938.272$ MeV/c^2, $M_n = 939.565$ MeV/c^2, $m_e = 0.511$ MeV/c^2]

Exercise 1.3.9

A nuclear reactor produces a total power of 2 GW. The fission reaction involved in the energy production is

$$^{235}_{92}U \rightarrow (A_1, Z) + (A_2, 92 - Z) + k\, n + \sim 200 \text{ MeV}$$

where k is an integer with values $2 \div 3$.

(a) Calculate the number of fission reactions per second.
(b) Knowing that the fissile nucleus, ${}^{235}_{92}$U, constitutes about 30% of the total fuel mass, estimate how much of this mass is consumed in a year.

Nuclear reactors are the major source of human-generated neutrinos. Neutrinos are produced in the β^- decay of the neutron-rich uranium fission fragments. This implies that they are actually $\bar{\nu}_e$. Assuming for simplicity that all neutrinos originate from the decay of ${}^{145}_{57}$La (as representative of all possible β^- decays) and the total decay rate corresponds to about 20% of the total fission rate,
(c) calculate the maximum neutrino energy;
(d) calculate the neutrino flux at 500 m from the reactor core.

(e) Knowing that the cross section for the reaction $\bar{\nu}_e + p \rightarrow n + e^+$ at these energies is about 6×10^{-44} cm^2, with an ideal detector of 1 ton active mass placed at 500 m from the core, how many neutrinos per year are detected?

Exercise 1.3.10

$^{27}_{14}$Si nuclei decay by β^+-decay into $^{27}_{13}$Al, whose binding energy is 224.95 MeV. The maximum kinetic energy of the emitted positrons is 3.79 MeV.

(a) Calculate the binding energy of the parent nucleus.
(b) Show that the difference in the binding energies between parent and daughter nuclei depend only on the Coulomb term of the *semi-empirical mass formula*.
(c) Estimate the $^{27}_{14}$Si radius, assuming a uniform charge distribution.

Exercise 1.3.11

The reaction $\nu_e + ^{71}_{31}$Ga $\rightarrow ^{71}_{32}$Ge $+ e^-$ is exploited for the detection of solar neutrinos in Gallium detectors [4, 5]. Knowing that the threshold energy is 233 keV and the $^{71}_{31}$Ga binding energy is 618.95 MeV, evaluate:

• the $^{71}_{32}$Ge binding energy;
• the relative error in the same binding energy, if the *semi-empirical mass formula* is used instead.

Assume the following numerical value $M_n - M_p - m_e = 782$ keV/c^2.

Exercise 1.3.12

According to certain theories the strengths of fundamental interactions can vary with time (on cosmological scale). For simplicity assume that only the electromagnetic coupling constant changes, the others remaining unchanged. The stablest nucleus for isobars having A $=$ 133 is nowadays $^{133}_{55}$Cs. If we assume that in another epoch it is $^{133}_{54}$Xe instead, estimate how much the electromagnetic coupling constant should change.
[$a_C = 0.697$ MeV, $a_A = 23.3$ MeV, $M_n - M_p - m_e = 0.782$ MeV/c^2]

Exercise 1.3.13

The interaction between photons and nuclei can proceed through the following processes (*nuclear photo-disintegration*)

$$\gamma + (A, Z) \rightarrow (A - 1, Z - 1) + p \qquad \gamma + (A, Z) \rightarrow (A - 1, Z) + n$$

In particular consider the photo-disintegration of $^{56}_{26}$Fe

$$\gamma + ^{56}_{26}\text{Fe} \rightarrow ^{55}_{26}\text{Fe} + n. \tag{1.5}$$

(a) Establish the photon threshold energy for reaction (1.5) in the case of an experiment with a photon beam against an iron fixed target.

(b) The same process (1.5) occurs when very energetic Cosmic Rays produced by an extra-galactic source propagate through the Universe. $^{56}_{26}$Fe nuclei involved in this process are ultra-relativistic and the hit photons belong to the *Cosmic Microwave Background* radiation. This radiation pervades the Universe and can be simplified by an isotropic flux of 1 meV energy photons.

Establish the threshold energy for $^{56}_{26}$Fe Cosmic Rays. Consider the case in which the photon direction is equal and opposite to the one of the propagating nuclei (*head-on collision*).

Exercise 1.3.14

In nuclear physics, the *separation energy* is the energy needed to remove one nucleon from a nucleus. It is denoted by S_p for a removed proton and S_n for a neutron.

(a) Derive an expression for $\Delta S = S_p - S_n$, using the *semi-empirical mass formula*.
(b) For light nuclei ($A < 40$) the stability line is approximately $A = 2Z$: write ΔS for this assumption and discuss the sign of the obtained expression.
(c) For heavy nuclei ($A > 100$) the stability line approaches $A = 2.5Z$: write ΔS and discuss the sign for this case.
(d) Calculate ΔS for the following (stable) nuclei $^{20}_{10}$Ne, $^{38}_{18}$Ar, $^{106}_{46}$Pd, $^{137}_{56}$Ba and $^{200}_{80}$Hg, discussing the results.

1.4   Nuclear Decays

Exercise 1.4.1

Consider the α-decay ^{240}Pu $\rightarrow ^{236}$U $+ \alpha$. Experimentally we observe two lines for the α kinetic energies, at 5.17 and 5.12 MeV. Which are the Q_α-values for the two modes? The lower energy line corresponds to the decay from an excited level ^{236}U*. Nucleus de-excites from this level via γ-decay. What is the γ energy?

Exercise 1.4.2

Consider the following decay chain

$$^{244}_{94}\text{Pu (81 Myr)} \ \rightarrow \ ^{240}_{92}\text{U (14 h)} \ \rightarrow \ ^{240}_{93}\text{Np (67 min)} \ \rightarrow \ ^{240}_{94}\text{Pu},$$

where half-lives are reported in brackets. Having 1 mol of pure ^{244}Pu, how many ^{240}U and ^{240}Np nuclei will be present after 1 month? Which is the radioactive process involved in each decay? What is the measured activity in the first decay process?

Exercise 1.4.3

The half life of ^{226}Ra is 1600 years. Assuming to have a pure 1 g source of ^{226}Ra, evaluate its activity.

Exercise 1.4.4

In dating organic specimens, carbon is usually used. Carbon-14 is a radioactive isotope of carbon that is produced by the action of Cosmic Rays on nitrogen in the atmosphere. If the flux of Cosmic Rays remains roughly constant over time, then the ratio of $^{14}_6C$ to the stable most abundant isotope $^{12}_6C$ reaches an equilibrium value of about 1.3×10^{-12}. $^{14}_6C$ decays by β^- with a half life of 5700 years.

Measuring the activity of a fossil of 5 g mass we get 3600 decays in 2 h. Estimate the age of the fossil.

Exercise 1.4.5

The activity of 1 g of ^{226}Ra is used to define the unity of activity of 1 Curie (Ci). The half life of ^{226}Ra is 1600 yr. Which is the mass of a ^{60}Co source ($T_{1/2} = 5.26$ yr), if we measure an activity of 10 Ci?

Exercise 1.4.6

Consider the following decay chain

- $N_1 \rightarrow N_2$, with a decay constant $\omega_1 = 10 \text{ s}^{-1}$;
- $N_2 \rightarrow N_3$, with $\omega_2 = 50 \text{ s}^{-1}$;
- N_3 is stable.

Assume that at time 0 the nuclei of type 1 are N_0 and none of the other types are present. Find the numbers N_1, N_2 and N_3 at any time. In particular, find the ratio N_3/N_1 after 1/4 s.

Exercise 1.4.7

^{238}U, the most abundant isotope of natural uranium, originates from the solidification of the Earth's crust occurred about 2.5 billions of years ago. Knowing that its half life is $4.5 \, 10^9$ years, derive

- the fraction of ^{238}U decayed so far;
- the specific activity of ^{238}U in Ci/g.

Exercise 1.4.8

Consider the following radioactive decays

1. $^{44}_{22}Ti \rightarrow ^{40}_{20}Ca + \alpha$
2. $^{241}_{95}Am \rightarrow ^{237}_{93}Np + \alpha$
3. $^{141}_{55}Cs \rightarrow ^{141}_{56}Ba + e^+ + \nu_e$
4. $^{69}_{28}Ni \rightarrow ^{69}_{29}Cu + e^- + \bar{\nu}_e$

Establish which ones are allowed or forbidden specifying the reason(s) in each case. [$B_\alpha = 28.3$ MeV, $M_p = 938.27$ MeV/c^2, $M_n = 939.57$ MeV/c^2, $m_e = 0.511$ MeV/c^2].

Exercise 1.4.9

Consider the following sequence of decays

$$^{79}_{38}Sr \ (2.25 \ min) \ \rightarrow \ ^{79}_{37}Rb \ (22.9 \ min) \ \rightarrow \ ^{79}_{36}Kr \ (35 \ h) \ \rightarrow \ ^{79}_{35}Br$$

where half-lives are reported in brackets and the last nucleus is stable.

Find the time at which an initially pure $^{79}_{38}Sr$ source has the maximum abundance of $^{79}_{37}Rb$ nuclei.

Exercise 1.4.10

A concrete basement (4 m $\times$ 5 m $\times$ 3 m) is not ventilated for long periods. Measuring the ^{222}Rn activity in the volume of the basement we get 100 Bq/m^3. Knowing that ^{222}Rn is produced along the ^{238}U sequential decay chain and this gas diffuses from the walls from a maximum depth of about 2 cm, find the ^{238}U concentration in the concrete (in number of ^{238}U nuclei per unit volume). The ^{238}U half-life is 4.5 billion years.

Exercise 1.4.11

The half-life of the ^{239}Pu decay ($^{239}Pu \rightarrow \ ^{235}U + \alpha$) has been measured immersing a 120 g source of ^{239}Pu in a liquid nitrogen vessel whose volume is large enough to contain the alpha decays. An evaporation rate of the liquid corresponding to a power of 0.231 W has been measured. Knowing that kinetic energy of alpha particles is 5.144 MeV, calculate the half-life of ^{239}Pu.

[Assume that the time of the measurement is much smaller than the half-life.]

1.5   Nuclear Models

Exercise 1.5.1

In a shell model the ground state energy is calculated using a Saxon-Woods potential, with parameters (R, d and V_0), tuned to fit experimental data. If in Nature the nuclear radius R were larger, e.g. 1.5 times the one found in experiments, all other model parameters being unchanged, discuss what we would expect for the energy of the ground state and the binding energy of the nucleus.

Exercise 1.5.2

On the basis of the shell model assign spin and parity, J^P, to the ground states of the following carbon isotopes: ^{11}C, ^{12}C, ^{13}C, ^{14}C.

Exercise 1.5.3

On the basis of the shell model assign spin and parity to the ground states of $^{33}_{16}S$, $^{39}_{19}K$ and $^{60}_{28}Ni$.

Exercise 1.5.4

Using the nuclear Fermi gas model, estimate the mean kinetic energy of nucleons for $^{16}_{8}O$ and $^{40}_{20}Ca$ nuclei.

Exercise 1.5.5

Using the shell model find J^P for $^{15}_{7}N$, $^{27}_{12}Mg$, $^{60}_{28}Ni$ and $^{87}_{38}Sr$, motivating the results in terms of their shell configurations.

Exercise 1.5.6

Consider the following oxygen isotopes $^{15}_{8}O$, $^{16}_{8}O$ and $^{17}_{8}O$ in their ground states.

1. Establish if they are stable. In case of decay, identify the possible decay type.
2. Using the shell model, assign spin and parity and evaluate the magnetic moment.

Exercise 1.5.7

Assign spin and parity to $^{7}_{3}Li$ and $^{29}_{14}Si$ ground states using

(a) the standard shell model;
(b) a shell model with *direct* spin-orbit interaction, that is having a *positive* term multiplying **L·S**.

Exercise 1.5.8

In the context of the shell model find the shell configurations and assign (whenever possible) spin and parity to the ground states of $^{51}_{24}Cr$, $^{52}_{24}Cr$ and $^{55}_{24}Cr$.
^{52}Cr is stable. Identify the possible decays of the other isotopes.

Exercise 1.5.9

The copper isotope $^{57}_{29}Cu_{28}$ decays by β^+ to $^{57}_{28}Ni_{29}$. For both nuclei involved in the decay,

(a) assign spin and parity to ground states and first excited levels;
(b) find the magnetic moments;
(c) estimate the maximum energy of the positron emitted in the decay. Show that it is possible to get such estimate without using the *semi-empirical mass formula* giving the motivation.

$[M_p = 938.27 \text{ MeV}/c^2, M_n = 939.57 \text{ MeV}/c^2, m_e = 0.511 \text{ MeV}/c^2]$

Exercise 1.5.10

Assign spin and parity to the following nuclei: $^{17}_{8}O$, $^{18}_{9}F$ and $^{207}_{82}Pb$.

References

1. Williams, W.S.C.: Nuclear and Particle Physics. Clarendon Press, Oxford (1991)
2. Chadwick, J.: Possible existence of a neutron. Nature **129**, 312 (1932)
3. Davis Jr., R., Harmer, D.S., Hoffman, K.C.: Search for neutrinos from the sun. Phys. Rev. Lett. **20**, 1205 (1968)
4. Anselmann, P., et al.: [GALLEX Collaboration], Solar neutrinos observed by GALLEX at Gran Sasso. Phys. Lett. B **285**, 376 (1992)
5. Abdurashitov, D.N., et al.: [SAGE Collaboration] Results from SAGE. Phys. Lett. B **328**, 234 (1994)

Chapter 2
Particle Physics

Abstract The problems collected in this chapter deal with Particle Physics from a phenomenological viewpoint. Whenever possible, the problems ask for numerical calculations, even though these are more properly estimates than actual calculations. This is because at introductory level the mathematical techniques are not yet available. In the first section, the general properties of the fundamental interactions are addressed. For this purpose, tools like Feynman diagrams and conservation properties are used. The further two sections are more specific in electro-weak and strong interactions.

2.1 Fundamental Interactions

Exercise 2.1.1

Show the Feynman diagrams at the lowest α order for the following processes

- $\gamma\gamma \to e^+e^-$
- $e^+e^- \to \mu^+\mu^-$
- $e^+e^- \to 4\gamma$

Exercise 2.1.2

The high energy neutrino-nucleon cross section can be written in natural units $(\hbar = c = 1)$ as

$$\sigma_{\nu N} = \frac{2G_F^2 s}{9\pi},$$

where G_F is the Fermi constant and s is the square total energy in the CMS. Find the neutrino energy above which the Earth becomes opaque to neutrinos. Assume that the Earth density is 2.15 g/cm^3 and its radius is 6000 km.

Exercise 2.1.3

Find the boson (or bosons) exchanged in the following processes

- $e^+ + e^- \to \mu^+ + \mu^-$

© Springer Nature Switzerland AG 2019
S. Petrera, *Problems and Solutions in Nuclear and Particle Physics*,
UNITEXT for Physics, https://doi.org/10.1007/978-3-030-19773-5_2

- $n \rightarrow p + e^- + \bar{\nu}_e$
- $\mu^- \rightarrow e^- + \bar{\nu}_e + \nu_\mu$
- $\nu_e + e^- \rightarrow \nu_e + e^-$
- $\nu_\mu + e^- \rightarrow \nu_\mu + e^-$

Exercise 2.1.4

Among the following processes, establish which ones are allowed or forbidden and which is the interaction type. Specify the reason(s) in each case.

- $\gamma + \gamma \rightarrow \gamma + \gamma$
- $e^+ + e^- \rightarrow 4\gamma$
- $p + \bar{p} \rightarrow W^- + X$
- $p + p \rightarrow K^+ + p$
- $\nu_\mu + e^- \rightarrow \nu_\mu + e^-$

Exercise 2.1.5

Among the following reactions, which ones are allowed? which ones are forbidden? Motivate the answers with Feynman diagrams, flavor flow diagrams or conservation principles.

- $e^+ + e^- \rightarrow \gamma + \gamma$
- $\pi^- + n \rightarrow K^- + \Lambda$
- $\Sigma^+ \rightarrow n + e^+ + \nu_e$
- $\Sigma^+ \rightarrow \Lambda + e^+ + \nu_e$
- $\rho^0 \rightarrow K^+ + K^-$
- $\bar{\nu}_e + e^- \rightarrow \bar{\nu}_e + e^-$
- $\nu_e + e^- \rightarrow \nu_e + e^-$

Exercise 2.1.6

In the following reactions X denotes an unknown particle. Identify the unknown particle, giving the motivation for your choice.

a. $\pi^- + p \rightarrow \Sigma^0 + X$
b. $e^+ + n \rightarrow p + X$
c. $\Xi^0 \rightarrow \Lambda + X$

Exercise 2.1.7

Draw the Feynman diagrams at the lowest order of the following processes:

(a) $e^+ + e^- \rightarrow e^+ + e^-$
(b) $e^+ + e^- \rightarrow \tau^+ + \tau^-$
(c) $\gamma + \gamma \rightarrow \gamma + \gamma$
(d) $p + \bar{p} \rightarrow W^- + X$
(e) $K^0 \rightarrow \pi^+ + \pi^-$

Exercise 2.1.8

For each of the following reactions establish whether it is allowed or not. If it is, establish the type of interaction and possibly draw the Feynman diagram. If it is not, specify the reason.

1. $\mu^+ \rightarrow e^+ + \gamma$
2. $e^- \rightarrow \nu_e + \gamma$
3. $p + p \rightarrow \Sigma^+ + K^+$
4. $e^+ + e^- \rightarrow \gamma$
5. $\nu_\mu + p \rightarrow \mu^+ + n$
6. $\nu_\mu + n \rightarrow \mu^- + p$
7. $e^+ + n \rightarrow p + \nu_e$
8. $e^- + p \rightarrow n + \nu_e$
9. $\pi^+ \rightarrow \pi^0 + e^+ + \nu_e$
10. $p + \bar{p} \rightarrow Z^0 + X$

2.2 Hadrons

Exercise 2.2.1

(1) An experiment is performed to study the inclusive K^0 production in the reaction

$$p + p \longrightarrow K^0 + X, \tag{2.1}$$

where X denotes any particle system (one or more particles).

– What are the values of its characteristic numbers (electric charge Q, B, S, ...)?
– What is the minimum number of particles in X?
– Propose a few solutions with known particles.

(2) With reference to the previous experiment, K^0-mesons are detected observing their decays. Knowing that: (a) $\tau(K_S^0) = 0.89 \times 10^{-10}$ s, (b) K^0-mesons produced in (2.1) have momenta in the range $1 \div 3$ GeV/c, (c) the main K^0 decay modes are $\pi^+\pi^-$ (69%) and $\pi^0\pi^0$ (31%),

– Propose and discuss a possible experimental set-up (detector types, sizes, positioning, etc.) aimed to observe reaction (2.1).

Exercise 2.2.2

Among the following reactions, establish which ones are allowed, motivating the answers

- $K^- + p \rightarrow \Omega^- + K^+ + K^0$
- $\psi \rightarrow \pi^+ + \pi^0 + \pi^-$

- $\pi^- + p \to \Sigma^+ + K^-$
- $\pi^- + p \to \pi^0 + \pi^0$
- $p + p \to n + \Delta^{++} + p + \bar{p}$

Exercise 2.2.3

Among the following decays, establish which ones are allowed and which are the interactions:

- $\phi \to \rho^0 + \pi^0$
- $\pi^0 \to e^+ + e^- + \gamma$
- $\Xi^- \to \Sigma^0 + \mu^- + \bar{\nu}_e$
- $\Sigma^- \to n + \pi^-$
- $\Xi^- \to \pi^0 + \pi^-$

[masses in MeV/c^2: $M_\phi = 1020$, $M_{\rho^0} = 769$, $M_{\pi^-} = 139.6$, $M_{\pi^0} = 135$, $M_{\Xi^-} = 1321$, $M_{\Sigma^0} = 1193$, $M_\mu = 105.6$, $M_{\Sigma^-} = 1197$, $M_n = 939.6$].

Exercise 2.2.4

Estimate the cross section $\sigma(e^+ + e^- \to \text{hadrons})$ in nb at CMS energy $\sqrt{s} = 2$ GeV, knowing that the cross section into $\mu^+ + \mu^-$ in natural units is:

$$\sigma = \frac{4\pi\alpha^2}{3s}$$

Exercise 2.2.5

The total decay width of the J/ψ is $\Gamma = 91$ keV. What is the mean lifetime? Which interaction is responsible for the decay?

Exercise 2.2.6

A 12 GeV/c π^+ beam is sent to a liquid hydrogen Bubble Chamber. An event is observed exhibiting an interaction with two charged tracks and two neutral vertexes pointing back to the interaction point. The two V^{0}s[1] have distances from the primary interaction point 37 cm and 11 cm respectively. The measurements for the first V^0 give $p_1^+ = 0.4$ GeV/c, $p_1^- = 1.9$ GeV/c and an opening angle $\theta_1 = 24.5°$ ($^+$, $^-$ stand for the sign of the particle charge). The second V^0 has $p_2^+ = 0.75$ GeV/c, $p_2^- = 0.25$ GeV/c and $\theta_2 = 22°$. The resolution on the invariant mass using the momentum errors is about 5%.

a. Which are the particles originating the two V^{0}s?;
b. give a possible interpretation of the observed reaction;
c. evaluate the lifetimes of the two particles.

---

[1]In the bubble chamber jargon, a "V^0" is a neutral particle decay into two (opposite) charged particles. Similarly a "trident" is a charged decay into three charged particles.

Exercise 2.2.7

Among the following reactions, which ones are allowed? which ones are forbidden? Explain why:

a. $\pi^- + p \to K^- + \Sigma^+$
b. $\pi^+ + p \to K^0 + \Sigma^+$
c. $\pi^- + p \to \Xi^- + K^+ + K^0$
d. $\Lambda \to \Sigma^- + \pi^+$
e. $K^- + p \to K^0 + n$
f. $\pi^+ + p \to \Lambda + K^+ + \pi^+$

Exercise 2.2.8

The Σ^0 strange baryon decays, with a mean lifetime $0.74 \ 10^{-19}$ s, as $\Sigma^0 \to \Lambda + \gamma$. It is an e.m. decay according to the lifetime value. The charged member of the Σ triplet, e.g. Σ^+, decays weakly in $0.80 \ 10^{-10}$ s.

a. Motivate the Σ^0 e.m. decay (why e.m.? why neither strong nor weak?)
b. Motivate the Σ^+ weak decay (why weak? why neither strong nor e.m.?)
c. Draw the Feynman graph(s) at the lowest order for the Σ^0 decay.

[masses: $\Sigma^0 = 1.193$ GeV/c^2, $\Sigma^+ = 1.189$ GeV/c^2, $\Lambda = 1.116$ GeV/c^2]

Exercise 2.2.9

The magnetic moment of the Λ-baryon has been measured in 1971 using nuclear emulsions [1]. Stacks of emulsion plates were placed at a distance of 10 cm from the target where Λ's were produced in the reaction

$$\pi^- + p \to \Lambda + K^0,$$

with π^- of 1 GeV/c momentum. In this strong interaction process Λ's are polarized with $P \approx 1$ along the normal to the scattering plane.

(a) Prove that the normal polarization is a consequence of the parity conservation in strong interactions. (*Hint: assume that a component of the spin is in the plane of the reaction and show that it violates parity conservation.*)

A magnetic field $B = 20$ Tesla, normal to the Λ flight path and parallel to the scattering plane, was pulsed at the arrival of each beam burst to the target (see Fig. 2.1). The Λ magnetic moment precesses about the direction of this magnetic field. Assume that the emulsion stack is positioned at a small angle ($\approx 0°$) with respect to the direction of the pion beam.

(b) Calculate the fraction of Λ-particles decaying before reaching the detector.
(c) Calculate the precession angle as Λ-particles reach the detector.

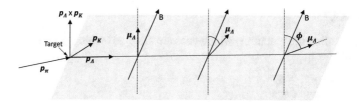

Fig. 2.1 Precession of the Λ magnetic moment in a uniform magnetic field

The Λ-decays into $p + \pi^-$ were observed in the emulsions and the angle θ^* of the proton with respect to the direction of the Λ magnetic moment was derived in the CMS. The θ^* distribution is

$$N(\cos\theta^*) \propto (1 - \alpha\cos\theta^*) . \qquad (2.2)$$

The decay asymmetry, f_+, is defined as the fractions of protons emitted forward $(\theta^* \leq 90°)$.

(d) Assuming $f_+ = 0.32$, calculate the value of the α-parameter in Eq. (2.2).
(e) What is the reason the observed asymmetry?

[$m_\Lambda = 1.116$ GeV/c^2, $\tau_\Lambda = 2.63 \ 10^{-10}$ s, $\mu_\Lambda = -0.61 \ \mu_N$, with $\mu_N = 1$ nuclear magneton (n.m.) = $3.15 \ 10^{-14}$ MeV/T; $m_p = 0.938$ GeV/c^2, $m_\pi = 0.140$ GeV/c^2, $m_{K^0} = 0.497$ GeV/c^2]

Exercise 2.2.10

An experiment is done to study of the associated production of strange particles

$$\pi^- + p \rightarrow \Lambda + K^0$$

with a π^- beam of momentum 1.5 GeV/c. Outgoing particles are analyzed in a magnetic spectrometer. An event is observed with two neutral vertexes (V^0), pointing back to the interaction point in the target, respectively with angles 58° and 21° with respect to the beam direction. The two V^{0}s exhibit the following features:

– The first V^0 is reconstructed unambiguously as a K^0 having a momentum 0.52 GeV/c;
– The second V^0, 10 cm far from the target, decays into a positive particle with $p_+ = 0.92$ GeV/c and a negative particle with $p_- = 0.21$ GeV/c. These particles are emitted at angles 4° (θ_+) and 14° (θ_-) respectively with respect to the V^0 direction.

a. Motivate why the second V^0 cannot be a Λ decaying as $\Lambda \rightarrow p + \pi^-$ (assume that the momentum resolution is 5%);
b. if instead the decay is $\Lambda \rightarrow p + e^- + \bar{\nu}_e$, evaluate the longitudinal momentum of the neutrino (i.e. w.r.t. the direction of the Λ);

c. if the particle is actually a Λ, calculate the lifetime.

[Particle masses in GeV/c^2: $p = 0.938$, $\pi^- = 0.140$, $\Lambda = 1.116$, $K^0 = 0.498$]

Exercise 2.2.11

Explain why each of the following particles cannot exist in the framework of the quark model:

(a) A baryon of spin 1.
(b) An antibaryon of electric charge +2.
(c) A meson with charge +1 and strangeness −1.

Exercise 2.2.12

What are the possible charges in the quark model for:

(a) a meson?
(b) an antibaryon?

2.3 Weak and Electro-Weak Interactions

Exercise 2.3.1

Calculate the length of an iron target a 300 GeV neutrino beam must cross in order that $1/10^9$ of the neutrinos interact. Assume that the high energy neutrino total cross-section is $\sigma_\nu = 10^{-38} E_\nu$ cm^2, with E_ν in GeV. The iron density is $\rho_{Fe} = 7.9$ g/cm^3.

Exercise 2.3.2

In the decay of D^0 ($= c\bar{u}$, $M_{D^0} = 1865$ MeV/c^2), two decay modes have the following measured ratio:

$$\frac{BR(D^0 \rightarrow K^- + e^+ + \nu_e)}{BR(D^0 \rightarrow \pi^- + e^+ + \nu_e)} = 11.37 \pm 0.05$$

Discuss and motivate this experimental result.
Hint: *consider both quark mixing and phase space contributions.*

Exercise 2.3.3

Draw the Feynman graphs associated to the following weak decays:

- $\Sigma^- \rightarrow n + e^- + \bar{\nu}_e$
- $\pi^+ \rightarrow \pi^0 + e^+ + \nu_e$
- $\tau^+ \rightarrow \pi^+ + \bar{\nu}_\tau$

Exercise 2.3.4

The neutron mean lifetime is 886 s. Assuming that the *Sargent rule* holds,

⋄ Estimate the Fermi constant.
⋄ Estimate the mean lifetime for the β^- decay of the nucleus $^{35}_{16}S$, knowing that its decay Q-value is 168 keV. For the calculation neglect the contributions of the nuclear transition amplitude and the Coulomb term.
⋄ Establish the nuclear spins of the parent and daughter nuclei in the context of the nuclear shell model.

N.B. The three answers are independent from each other.

Exercise 2.3.5

The OPERA experiment [2] at Gran Sasso Lab has detected a few events interpreted as charged current (CC) interactions of tau neutrinos. The detector was exposed to a muon neutrino beam, produced at CERN (Long Baseline Neutrino Beam), and having an average energy of 20 GeV. The observation of interactions $\nu_\tau \to \tau^-$, followed by the decays of the τ leptons, has been interpreted as evidence of the oscillation in flight $\nu_\mu \to \nu_\tau$. Using the oscillation parameters obtained in other experiments, the oscillation probability $P(\nu_\mu \to \nu_\tau)$ at this distance and energy is expected to be about 1.5%.

The primary background of this experiment is due to charm production in CC muon neutrino interactions ($\nu_\mu \to \mu^- + c$), because the mean lifetimes of charmed hadrons (around $10^{-12} \div 10^{-13}$ s) are comparable to the tau lifetime and they can mimic the tau decay.

1. What is the expected fraction of ν_μ CC interactions producing charm to all CC interactions?
2. What is the signal-to-noise ratio: $(\nu_\tau \to \tau^-)/(\nu_\mu \to \mu^- + c)$?
3. Show a few possible τ^- decay modes and draw the associated Feynman graphs.

Exercise 2.3.6

The branching ratios of the following Σ^- decays are

$$BR(\Sigma^- \to n + e^- + \bar{\nu}_e) = 1.02 \ 10^{-3}$$

$$BR(\Sigma^- \to \Lambda + e^- + \bar{\nu}_e) = 0.57 \ 10^{-4}$$

Draw the Feynman diagrams of these decays and estimate the Cabibbo angle from their ratio.
$[m_{\Sigma^-} \simeq 1197 \text{ MeV/}c^2, m_\Lambda \simeq 1116 \text{ MeV/}c^2, m_n \simeq 940 \text{ MeV/}c^2]$

Exercise 2.3.7

Consider the following decay rates:

$$\Gamma(D^+ \to \bar{K}^0 + e^+ + \nu_e) = 7 \times 10^{10} \, \text{s}^{-1} \qquad \Gamma(\mu^+ \to e^+ + \nu_e + \bar{\nu}_\mu) = \frac{1}{2.2 \, \mu\text{s}}$$

Motivate the ratio between these two values.
[$m_{D^+} = 1870 \, \text{MeV/c}^2$, $m_{\bar{K}^0} = 498 \, \text{MeV/c}^2$, $m_\mu = 106 \, \text{MeV/c}^2$, $m_e = 0.5 \, \text{MeV/c}^2$]

Exercise 2.3.8

Atmospheric neutrinos are produced by the interaction of cosmic rays in the Earth atmosphere. They emerge from the decays of charged pions, which populate the cascades produced in the atmosphere, and the following decays of muons. What is the expected ratio between muon and electron neutrinos $(\nu_\mu + \bar{\nu}_\mu)/(\nu_e + \bar{\nu}_e)$?
[For a comprehensive answer one should consider that the bulk of the atmospheric neutrinos have energies between 0.1 and 1 GeV and the mean lifetimes of charged pions and muons are $2.6 \, 10^{-8}$ and $2.2 \, 10^{-6}$ s respectively.]

Exercise 2.3.9

Consider the decays $\mu^- \to e^- + \bar{\nu}_e + \nu_\mu$ and $\tau^- \to e^- + \bar{\nu}_e + \nu_\tau$. The branching ratios are 100% for the former and 18% for the latter. The muon mean lifetime is 2.2 μs. Calculate the tau mean lifetime.
[Masses: $\mu = 106 \, \text{MeV/c}^2$, $\tau = 1777 \, \text{MeV/c}^2$.]

References

1. Dahl-Jensen, E., et al.: A new measurement of the magnetic moment of the Λ-hyperon. Nuovo Cim. A **3**, 1 (1971)
2. Agafonova, N., et al.: [OPERA Collaboration], Observation of a first ν_τ candidate in the OPERA experiment in the CNGS beam. Phys. Lett. B **691**, 138 (2010)

Chapter 3
Experiments and Detection Methods

Abstract The subject of this chapter is related to experiments and detection methods. It is divided into three sections: relativistic kinematics, passage of particles and radiation through matter and detection techniques and methods. In the first section several problems involving relativistic scattering and decay are proposed. Here we use the relativistic invariant approach (and the formalism) following Relativistic Kinematics, by R. Hagedorn [1]. The interaction of radiation with matter deals with the electromagnetic processes occurring to particles at each experimental site and which also make them detectable. An excellent review of this subject can be found in [2]. In the last section problems address experimental and detection techniques in realistic and actual cases.

3.1 Kinematics

Exercise 3.1.1

An experiment is done to study the Σ^+ decay. To this aim the used detector is a "tracker" (i.e., a detector of ionising particles having a high spatial resolution, e.g. a bubble chamber, a drift chamber, etc.). Σ^+-baryons are produced in the reaction

$$\pi^+ + p \rightarrow \Sigma^+ + K^+$$

($m_\pi = 0.1396$ GeV/c^2, $m_p = 0.9383$ GeV/c^2, $m_\Sigma = 1.189$ GeV/c^2, $m_K = 0.4937$ GeV/c^2), from a 20 GeV/c π^+ beam hitting a thin target. The detector can be simplified as a cylinder, with radius R and length L with the axis coincident to the beam line and placed immediately downstream of the target.

For simplicity let us assume that all Σ^+'s decay within three times their mean lifetime ($\tau_\Sigma = 0.799 \times 10^{-10}$ s).

1. Is the designed apparatus capable of detecting all the Σ^+'s produced?
2. What is the minimum detector length to contain all the Σ^+-decay points?
3. What is the minimum detector radius to fulfill the same requirement?
4. Is the designed apparatus capable of detecting all the K^+'s produced?

© Springer Nature Switzerland AG 2019   23
S. Petrera, *Problems and Solutions in Nuclear and Particle Physics*,
UNITEXT for Physics, https://doi.org/10.1007/978-3-030-19773-5_3

5. If it is not the case, which is the fraction of detectable kaons, assuming that they are produced isotropically in the CMS?

Exercise 3.1.2

(1) Consider a pion beam with total energy E_π impinging on a hydrogen target to produce a resonance having mass M. The resonance fastly decays in two particles with masses m_1 and m_2. Knowing that $M = 2.58\, m_1$ and m_2 is negligible with respect to m_1, establish the minimum value of E_π to have a maximum production angle for particle 1.

(2) The produced resonance is $\Delta(2420)$, where the number in brackets stands for the invariant mass in MeV. A possible decay channel, different from the one considered in (1), is $\Delta(2420) \rightarrow \Sigma + K$ [1], with $m_\Sigma = 1.189\,\text{GeV/c}^2$ and $m_K = 0.494\,\text{GeV/c}^2$. Assume that the beam energy is that determined in (1). If a Σ is emitted at an angle of $120°$ in the CMS, (a) what is the corresponding angle in the Laboratory system? (b) what is its momentum?

(3) An experimental set-up has been designed to detect the Σ's produced in the above reaction. A detector length of 26 cm fulfills the requirement that at least 99% of the Σ-decay points are contained in the detector. What is the Σ mean lifetime?

Exercise 3.1.3

A negative pion beam is incident on a proton target to produce the reaction $\pi^- + p \rightarrow \Lambda + K^0$.

1. Calculate the minimum pion energy for which the reaction is allowed.
2. Setting the pion energy at $E_\pi = 2\,\text{GeV}$, establish if there is a maximum production angle (in the LS) for the Λ-particle.

$[M_\pi = 140\ \text{MeV/c}^2,\ M_p = 938\ \text{MeV/c}^2,\ M_\Lambda = 1116\ \text{MeV/c}^2,\ M_K = 498\ \text{MeV/c}^2]$

Exercise 3.1.4

In a certain experiment neutral particles having energy 10 GeV are observed to decay into $\pi^+ + \pi^-$. The opening angle distribution exhibits a minimum value of about $5.2°$. Calculate the mass of the particle.

Exercise 3.1.5

In an electron-electron collider two e^- beams hit each other in opposite directions. The respective beam energies are $E_1 = 12\,\text{GeV}$ and $E_2 = 5\,\text{GeV}$.

- What is the total CMS energy?
- What are the electron momenta in the CMS?
- What are the beta and gamma, β_{CM} and γ_{CM}, of the Lorentz transformation LS $\rightarrow$ CMS?
- In a collider with equal beam energies, $E_1 = E_2$, establish the relation between CMS and LS.

---

[1] Here charges are omitted: Δ can be $++, +, 0$ and $-$, Σ and K can be $+, 0$ and $-$.

Exercise 3.1.6

Consider the weak decay $\Xi^0 \rightarrow \Sigma^+ + e^- + \bar{\nu}_e$. Assuming that the Ξ^0-baryon ($M_{\Xi^0} = 1315$ MeV/c^2) is at rest, calculate the maximum and minimum energies of the electron.
[$M_{\Sigma^+} = 1189$ MeV/c^2, $M_{e^-} = 0.511$ MeV/c^2]

Exercise 3.1.7

A negative pion beam with momentum $p_\pi = 20$ GeV/c is incident on a proton target to produce the reaction

$$\pi^- + p \rightarrow \Sigma_c^0 + \bar{D}^0.$$

Considering a $\bar{D}^0$ produced at the maximum angle in the LS and decaying into $\pi^+ + \pi^-$, calculate the minimum opening angle between the two pions.
[$M_\pi = 0.14$ GeV/c^2, $M_p = 0.94$ GeV/c^2, $M_D = 1.86$ GeV/c^2, $M_\Sigma = 2.45$ GeV/c^2]

Exercise 3.1.8

In the study of Ultra High Energy Cosmic Rays (UHE stands for $E_{CR} > 10^{18}$ eV) the following process (called the "GZK effect" [3]) occurs

$$p + \gamma_{CMB} \rightarrow p + \pi^0. \tag{3.1}$$

This reaction represents the photo-production of pions induced by CR protons as they cross the Universe and interact with the photon background contained therein. These photons, which are usually referred to as *Cosmic Microwave Background* photons (and are denoted by γ_{CMB}), represent the residual radiation emitted after the Big Bang.

- Show the dependence of the threshold energy of reaction (3.1) as a function of the scattering angle between the proton and the CMB photon.
- Calculate the minimum energy for a CR proton to make pion photo-production.

Use the following numerical values: $E_{\gamma CMB} = 10^{-3}$ eV, $M_p = 0.94$ GeV/c^2, $M_\pi = 135$ MeV/c^2.

Exercise 3.1.9

Consider the following neutron capture reaction, $n + p \rightarrow d + \gamma$, assuming that the initial particles are at rest. From a measurement of the photon energies we get $E_\gamma = 2.230 \pm 0.005$ MeV. Calculate the deuteron mass and its error.
[$M_p = 938.272$ MeV/c^2, $M_n = 939.565$ MeV/c^2]

Exercise 3.1.10

1. In the annihilation of anti-protons with momentum $p_{\bar{p}} = 1$ GeV/c against protons at rest $K^- K^+$ pairs are produced. Consider the case of symmetric production (kaons having the same energies), corresponding to kaons emitted at 90° with respect to the anti-proton direction in the center-of-momentum system. What are the momenta and angles of the kaons in the LS?

2. We want to detect the produced kaons using two ionization detectors placed at the angles determined above. Assume that the detectors are 10 cm thick and contain a gas with density $\rho = 2$ mg/cm^3 and ionization potential $I = 15$ eV. If the efficiency for producing electron-ion pairs is $\epsilon_p = 20\%$ and the collection efficiency is $\epsilon_c = 30\%$, what is the number of pairs collected in each detector?
[$M_p = 0.94$ GeV/c^2, $M_K = 0.49$ GeV/c^2]

Exercise 3.1.11

In a pp experiment an event is observed with two opposite sign muons ($m_\mu = 106$ MeV/c^2), emitted along opposite directions, having momenta respectively 45 MeV/c and 30 GeV/c. If these muons are originating from the decay of a particle, what are the momentum and mass of this particle?

Exercise 3.1.12

Ultra High Energy Cosmic Rays propagating through the Universe undergo the following reaction

$$p + \gamma_{CMB} \to p + e^+ + e^-.$$

This process originates from their collisions against the *Cosmic Microwave Background* photons (γ_{CMB}). These photons pervade the whole Universe and are distributed isotropically.

Calculate the energy threshold as a function of the scattering angle and determine the angle corresponding to the minimum threshold value. Assume $E_{\gamma_{CMB}} \simeq 1$ meV.

Exercise 3.1.13

In the Large Hadron Collider (LHC) at CERN, Geneva, proton-proton interactions are investigated at the highest energy ever reached by humans. Protons hit each others with opposite momenta so that the laboratory is in the CMS. The current total energy is 13 TeV.

(a) Cosmic Rays hitting the Earth are mostly protons having energies between approximately 10^8 and 10^{20} eV. What is the energy in eV to produce pp interactions equivalent to the ones studied at LHC?
(b) Which velocity should have an insect ($M \approx 0.25$ g) to have the same kinetic energy of these cosmic rays?

Exercise 3.1.14

The *rapidity* is a quantity that is used in high energy hadronic interactions, defined as

$$y = \frac{1}{2} \ln \frac{E + p_\parallel}{E - p_\parallel},$$

where E is the particle energy and $p_\parallel$ is the parallel component of its momentum. In the pp scattering at colliders, this component is identified as the projection onto the beam directions in the interaction point.

(a) Show that under Lorentz transformation, the rapidity is only shifted by a quantity that depends on the β of the transformation.
(b) Calculate the maximum and minimum rapidities in the case of LHC (pp colliding beams at 13 TeV in the center-of-momentum reference system).
(c) Show that in the ultra-relativistic limit ($E \simeq p$) the rapidity can be approximated by the *pseudorapidity*, defined as

$$\eta = -\ln \tan \frac{\theta}{2},$$

where θ is the particle angle ($p_{\parallel} = p \cos \theta$).
(d) Compare the rapidities and the pseudorapidities at $90°$ and $1°$ in the LHC case.

Exercise 3.1.15

The so called "impact parameter" method [4] is used to select particle decays with very short lifetimes. This method is based on the idea that, if the observed event has a track (or more tracks) coming from a secondary decay, the backward prolongation of this track does not point to the vertex of the primary interaction. This method is particularly interesting for those cases in which the track of the decaying particle is too short to be observed. Defining the impact parameter as the distance between the line corresponding to the decay track and the primary vertex, we have $\Delta = L \sin \theta$ where L is the decay distance and θ is the angle of the particle emitted in the decay (see figure).

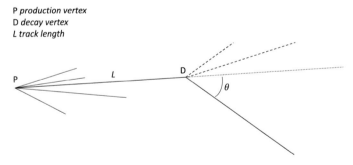

P production vertex
D decay vertex
L track length

- Prove that for ultra-relativistic particles (primaries and secondaries), the impact parameter does not depend on the momentum of the decaying particle, which is not the case for the decay length.
  (*Hint*: calculate the expression $\beta\gamma \sin \theta$ as a function of the $\theta^\star$ in the CMS.)
- In these conditions, we can write that $\Delta = ct \, \tan(\theta^\star/2)$, where t is the particle lifetime and $\theta^\star$ is the angle of the emitted particle in the CMS. Consider the D^+-decay and calculate, for a lifetime equal to the mean lifetime ($\tau = 1.04 \times 10^{-12}$ s), the mean impact parameter and the one corresponding to $\theta^\star = 90°$.

Exercise 3.1.16

In an accelerator experiment two opposite sign muons are interpreted as being due to the decay of a neutral particle. The muons ($m_\mu = 106$ MeV/c^2) are emitted with an opening angle of $42°$ and momenta respectively 7.4 and 2.6 GeV/c. Calculate the mass and the energy of the neutral particle.

During the same experiment several events are observed with roughly the same energy and decaying particles: calculate the energy of the muons when the opening angle is minimum.

Exercise 3.1.17

High Energy neutrino beams are produced in proton accelerator sites, injecting almost mono-energetic secondary pion beams (e.g. π^+) into a long vacuum pipe to allow their decays ($\pi^+ \to \mu^+ + \nu_\mu$).

(a) Find the neutrino energy in the pion rest frame (RF).

In the laboratory system the neutrino energy depends on the decay angle. Assume that the pion beam energy is 200 GeV.

(b) What is the maximum neutrino energy in the LS?
(c) What is the neutrino energy for neutrinos emitted in the forward hemisphere ($\theta_\nu^* \le 90°$) in the pion RF?
(d) What is the maximum LS angle for the neutrinos emitted forward in the pion RF?

[Particle masses: $\pi^+ = 0.140$ GeV/c^2, $\mu^+ = 0.106$ GeV/c^2.]

Exercise 3.1.18

An astrophysical source at 5000 light-years distance emits neutrons. What is the minimum energy that neutrons must have to reach the Earth?

To make the estimate, assume that neutrons decay after a mean lifetime and the half-life is about 10 min.

Exercise 3.1.19

The Universe is filled with black-body microwave (*CMB, Cosmic Microwave Background*) radiation. The average photon energy is $E \approx 10^{-3}$ eV. Very high energy photons from astrophysical sources make electron-positron pairs in their collisions with *CMB* photons.

(a) Draw the Feynman graph for this process and evaluate the α order of the cross section.
(b) What is the minimum photon energy to produce pairs, in the case of head-on collisions?
(c) For the same case, find the Lorentz factor γ of the CMS reference system.

3.2 Interaction of Radiation with Matter

Exercise 3.2.1

The total photon absorption coefficient for 5 MeV photons in lead is about 0.04 cm^2/g. Knowing that the density is 11.3 g/cm^3, what is the Pb thickness to halve the intensity of a 5 MeV photon beam? What is the thickness to allow a 5% beam survival?

Exercise 3.2.2

The radiation length of lead ($\langle A \rangle = 207$, $\rho = 11.3$ g/cm^3) is 5.6 mm. What is the absorption coefficient and the cross section for e^+e^- pair production from high energy photons?

Exercise 3.2.3

A 20 GeV/c muon crosses a 50 cm thick slab of magnetized iron, where a field $B = 2$ Tesla is produced parallel to the plane of the slab. The initial direction of the muon is normal to the slab. Knowing that the iron radiation length and density are $X_0 = 1.8$ cm and $\rho = 7.87$ g/cm^3 respectively, find the magnetic deflection angle, the momentum at the exit of the slab and the multiple scattering dispersion in the plane containing the muon trajectory.

Exercise 3.2.4

Find the Compton scattering mean pathlength in water ($\langle Z/A \rangle = 0.56$) for 1 keV photons.

Exercise 3.2.5

A thin X-ray beam is sent to an imaging detector exposed to a magnetic field, $B = 0.1$ T, uniform and with a direction normal to the beam. A Compton event is identified with an electron ejected at an angle $\phi = 10°$, with respect to the beam direction. The electron generates a circular track and leaves the detector at a distance $L = 3$ cm from the initial point. A sagitta $s = 0.2$ cm is measured from the electron trajectory. Calculate the energy of the beam and that of the scattered photon.

Exercise 3.2.6

A 20 GeV π^- beam with a current intensity of 10 μA is monitored by an ionization counter. This counter can be viewed as a gas cell having thickness 1 cm, density 1.8 10^{-3} g/cm^3 and mean ionization potential $\langle I \rangle = 15$ eV. Assuming that each electron-ion pair created in the ionization process is actually detected, estimate the current measured in the detector.

Exercise 3.2.7

An extracted π^+ beam contains a minor contamination of protons: both particles have the same momentum $p = 5$ GeV/c. To separate the two beam components two

Cherenkov detectors, having refractive indexes $n_1 = 1.05$ and n_2, are placed along the beam line. Find a possible choice for n_2 to achieve the beam separation. [$M_\pi = 0.139$ GeV/c^2, $M_p = 0.938$ GeV/c^2]

Exercise 3.2.8

A 500 MeV/c muon beam (mass = 0.106 GeV/c^2) is incident normally on a copper slab ($\rho = 9$ g/cm^3, $X_0 = 1.4$ cm).

(a) Find the thickness needed to stop the beam.
(b) If instead the slab is $d = 10$ cm thick, calculate the energy and the multiple scattering angle of the muons after the slab.

Exercise 3.2.9

To study the photoelectric effect a monochromatic UV beam ($\lambda = 200$ nm) is sent to a silver foil. Knowing that the electron binding energy in silver is $W = 4.73$ eV, establish if the photoelectric process actually occurs and, if so, find the kinetic energy of emitted electrons.

Exercise 3.2.10

A cosmic photon having energy 100 GeV interacts with air molecules hitting the Earth atmosphere. Knowing that in air the critical energy is 80 MeV and the radiation length is 37 g/cm^2, estimate the thickness of crossed atmosphere (in g/cm^2) where the electromagnetic shower is developed at its maximum.

Exercise 3.2.11

In the *Positron Emission Tomography* (PET), the e^+e^- annihilation is exploited to produce photons which are detected and measured in the apparatus. The positrons are produced in the organic material where a β^+ emitting radionuclide is introduced for this purpose. Subsequently e^+e^- annihilate at rest into pairs of photons. Assuming that photons are detected through the Compton scattered electrons, find their minimum and maximum energy.

Exercise 3.2.12

The Earth's atmosphere has a total thickness of 1030 g/cm^2 at the sea level.

• Estimate the minimum energy for a vertical muon ($m_\mu = 106$ MeV/c^2) to cross the whole atmosphere.
• Knowing that the air mean ionization potential is about 10 eV, find the mean number of electrons extracted during the muon path to the ground.

Exercise 3.2.13

A 1 GeV electron beam crosses normally a lead plate having thickness $X_0/20$, where X_0 is the lead radiation length. Establish which of the following processes (i) bremsstrahlung, (ii) multiple scattering dominates the angular distribution.

Exercise 3.2.14

A thin 3 GeV muon beam hits a copper slab 10 cm thick ($\rho = 9$ g/cm³, $X_0 = 13.3$ g/cm²). Evaluate the energy loss and the broadening of the beam produced by multiple scattering.

Exercise 3.2.15

Explain in few lines, suggesting an example, how the knowledge of the proton range can be used to infer the range of another particle with different charge and mass.

Exercise 3.2.16

Estimate the mean energy radiated from 1 GeV electrons crossing an aluminium plate 5 cm thick.
[$A = 27$, $Z = 13$, $\rho = 2.7$ g/cm³, $D = 4N_A\alpha r_0^2 = 1.4 \times 10^{-3}$ cm²/g]

Exercise 3.2.17

A muon beam with momentum $p = 500$ MeV/c enters a region with an uniform magnetic field $B = 0.1$ T, orthogonal to the beam direction. Due to the magnetic field the beam is making a curved orbit.

(a) Calculate the radius of the orbit in vacuum.
(b) Assuming that particles are moving in a gas ($\rho = 2\ 10^{-3}$ g/cm³), calculate the radius of curvature after a complete round.

Exercise 3.2.18

In Compton scattering electrons show a peak in their distribution at a characteristic maximum energy ('Compton edge'). What is the value of this energy for 0.5 MeV photons?

Exercise 3.2.19

An ionization-sensitive detector, exposed to a γ source, measures the energy spectrum shown in the figure.

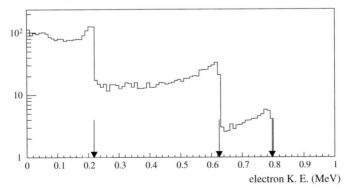

The spectrum is produced by Compton scattered electrons that release their kinetic energy in the detector and is interpreted as originated by three γ-decay lines whose corresponding 'Compton edges' are represented by arrows in the figure. Find the Q_γ values for the decays.

Exercise 3.2.20

High energy muons are produced in extensive air showers as Cosmic Rays hit the Earth atmosphere. Assume for simplicity that all muons have energy 10 GeV and their production occurs at 10 km a.s.l. Answer the following questions:

(a) knowing that the air refractive index is $n = 1.00029$, do such muons produce Cherenkov photons?
(b) if yes, what is the opening angle (w.r.t to the muon direction) of the produced photons?
(c) how many photons hit the sea level?

Exercise 3.2.21

A detector is used to measure the energies of photons from a monoenergetic source. The photons are collimated and hit the detector inclined at 30° with respect to the beam, as shown in the figure. The detector measures Compton scattered electrons and consists in a bundle of scintillating fibres (a single fiber is shown in the figure) with a photomultiplier applied to their exit. Assume that the fibre has a thickness $d = 2$ mm, density $\rho = 1$ g/cm^3, $Z/A \simeq 0.5$ and an acceptance of 15°. The angular acceptance is meant as the semi-opening angle with respect to the bundle axis for which the electron energy release is fully contained. In the detector a release of 2 MeV is measured as due to the scattered electrons. Calculate:

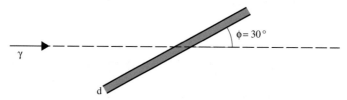

(a) the beam photon energy;
(b) the cross section for the electrons accepted by the detector;
(c) the fraction of detected electrons.

Exercise 3.2.22

After crossing one radiation length, what is the mean energy lost by 1 GeV electrons?

3.3 Detection Techniques and Experimental Methods

Exercise 3.3.1

Two equal scintillation counters, S1 and S2, 1 cm thick, placed on a beamline at a relative distance $L = 3$ m, are used to measure the time-of-flight of crossing particles. All the particles in the beamline have the same momentum p. The scintillator radiation length is $X_o = 40$ cm.

(1) Show that, in the limit $E \gg m$, the following relation holds for the difference between the time-of-flights of two particles having square mass difference $\Delta m^2 = m_1^2 - m_2^2$

$$\Delta T = \frac{L}{2c} \frac{\Delta m^2}{p^2} ,$$

where $\Delta T = T_2 - T_1$ and T_i is the time-of-flight of particle i.

(2) If this set-up is used to separate π^+ ($m_\pi = 139$ MeV/c^2) from K^+ ($m_K = 493$ MeV/c^2) in a beamline with $p = 1$ GeV/c, what is the time resolution needed to achieve a particle discrimination within 4 standard deviations?

(3) Assume that S1 and S2 are segmented in parallel strips 5 cm wide and a third identical detector S3 is inserted in between S1 and S2 at the same distance from both. A uniform magnetic field of 1 Tesla, normal to the plane of the figure, is set up in the region between S1 and S2. This system is used as a spectrometer to get the momentum from a sagitta measurement.

(a) Show that the multiple scattering does not affect the measurement.
(b) Evaluate the momentum resolution $\Delta p / p$, for 1 GeV/c pions.

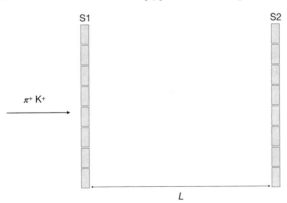

Exercise 3.3.2

A 300 MeV/c muon enters a magnetic field region, whose horizontal section is shown in the figure. Assume that the magnetic field is uniform, is normal to the plane of the figure and has an intensity of 0.5 Tesla. The initial muon direction is normal to the magnetic field direction. At the time t_0, when the muon is at O in the figure, the magnetic field is switched on to keep it on a circular orbit of radius R. The medium crossed by the muon in its trajectory has a density $\rho = 10^{-3}$ g/cm^3. Travelling along its trajectory the muon crosses two 2 mm thick iron septa ($\rho_{Fe} = 7.87$ g/cm^3).

(a) Estimate the difference $\Delta B = B' - B$, where B is the magnetic field intensity at the time t_0 and B' is the field needed to keep the muon in the circular orbit after one turn.
(b) If the same apparatus is operated in the vacuum and the iron septa are removed, what is the mean number of turns made by the muon before its decay?

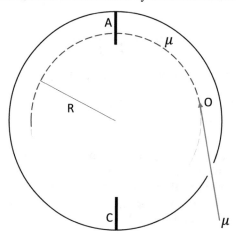

Exercise 3.3.3

The *SuperKamiokande* detector [5] consists in a huge vertical cylinder filled with pure water ($n = 1.33$, $\rho = 1$ g/cm^3) looked by a grid of photodetectors. Muonic atmospheric neutrinos are detected through the Cherenkov radiation emitted by the muons produced in the neutrino interactions with water nuclei. Assume for simplicity that all muons have a momentum of 1 GeV/c.

1. Estimate the maximum pathlength of the muons.
2. What is the fraction of this pathlength for which muons emit Cherenkov radiation?
3. What is the radius of the circle in the bottom of the detector illuminated by the Cherenkov radiation, if a muon is produced at 50 cm from the cylinder base, directed downward along its axis (see figure)?

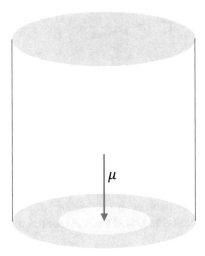

Exercise 3.3.4

In a colliding beam accelerator electrons and positrons make interactions (with equal and opposite momenta) at a total energy of 90 GeV. An annular detector at a distance of 2 m from the interaction point measures the rate of particles produced in the interaction $e^+ + e^- \rightarrow e^+ + e^-$. Assume that the detector has internal and external diameters 12 and 20 cm respectively and a negligible thickness. The detected rate turns out to be 1 event-per-second. Knowing that the Bhabha cross section at small angles can be written as

$$\frac{d\sigma}{d\theta} = \frac{8\pi\alpha^2}{E_e^2} \frac{(\hbar c)^2}{\theta^3},$$

where E_e is the energy of e^- and e^+, calculate the collider luminosity.

Exercise 3.3.5

Theories unificating strong and electro-weak interactions *(Grand Unified Theories)* predict that nucleons are unstable and decay with a mean lifetime around 10^{32} years. Estimate the minimum mass of a nucleon-decay experiment to detect at least one decay per year.

Exercise 3.3.6

A charged pion beam having momenta ranging between 0.5 and 1.5 GeV/c is collimated by a narrow slit 1 cm wide after having crossed a region 1.1 m long, where a magnet generates a 0.2 Tesla uniform field (see figure). What is the distance between the slit and the end of the magnet to select a momentum $p_0 = 1$ GeV/c $\pm 5\%$.

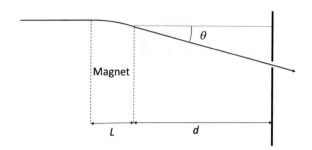

Exercise 3.3.7

A muon beam is circulating in an accumulation ring of 14 m radius with a magnetic field of 0.5 Tesla. Knowing that $m_\mu = 106$ MeV/c^2 and $\tau_\mu = 2.2$ μs, what are the muon momentum, the revolution period and the fraction of muons lost in a turn?

Exercise 3.3.8

An experiment is made to search for proton decays as predicted by *Grand Unified Theories (GUT)*. The detector consists of a huge water container of cubic shape where decays can be observed through the Cherenkov radiation emitted by the proton decay products. In particular the search addresses the observation of the 'golden' channel

$$p \rightarrow e^+ + \pi^0$$

Positrons and photons (emitted in the decay $\pi^0 \rightarrow \gamma\gamma$) induce e.m. cascades in the detector and the charged particles contained therein emit Cherenkov photons.

1. Estimate the size of the detector required to fully contain the e.m. cascades.
2. Assuming that the Cherenkov photon yield in the detection wavelength interval is about $I_0 = 400$ cm^{-1}, estimate the total number of photons.

[Water: radiation length $X_0 \simeq 36$ g/cm^2, critical energy $E_c \simeq 80$ MeV; proton mass $m_p \simeq 0.94$ GeV/c^2, positron mass $m_e \simeq 0.511$ MeV/c^2, neutral pion mass $m_{\pi^0} \simeq 0.135$ GeV/c^2]

Hint: having in mind that the photon opening angle in the π^0 decay is negligible, the decays appears as two back-to-back cascades.

Exercise 3.3.9

A 300 GeV proton beam circulates in an accumulation ring where vacuum is kept at 10^{-11} atm. Protons interact against the residual air molecules (assume for air: $Z = 7$, $A = 14$, $\rho = 1.25 \times 10^{-3}$ g/cm^3 in standard atmospheric conditions). Knowing that the total p-Air cross section is about 300 mb, calculate the mean beam lifetime.

Exercise 3.3.10

A hadron beam is incident on a lead (density 11.3 g/cm^3) target 2 mm thick. The beam has a circular section of radius 1 cm. Assuming that the total cross section is 30 mb,

- What is the number of scattering centres within the beam area?
- What is the fraction of the beam scattered by the target?

Exercise 3.3.11

The Earth is hit continuously by solar neutrinos (i.e. produced by the Sun). They have an energy spectrum extending up to about 10 MeV. Neutrinos above 4 MeV are detected in a huge detector containing 50,000 tons of water, observing their interaction against atomic electrons. At this energy the $\nu_e e^-$ cross section is about $7 \, 10^{-20}$ b and the neutrino flux at the Earth is about 10^6 cm^{-2}s^{-1}. Calculate the number of neutrino interactions per year.

Exercise 3.3.12

Consider cosmic photons of energy 500 GeV hitting the Earth atmosphere. We use a ground apparatus to detect the electromagnetic showers produced in the atmosphere.

(a) Knowing that in air the critical energy is $E_c^{atm} = 80$ MeV and the radiation length is $X_0^{atm} = 37$ g/cm^2, find the optimal altitude to measure the cascades at their maximum development for vertical photons. [Assume that the vertical grammage (the atmosphere depth in g/cm^2) depends on altitude as $X_v(h) = X_v(0) \exp(-h/h_0)$, where con $X_v(0) = 1000$ g/cm^2 and $h_0 = 7$ km.]
(b) Assume that the detector is a water tank viewed by optical sensors. Can we exploit the Cherenkov effect to detect the e.m. shower? [$n_{H_2O} = 1.33$]
(c) The critical energy in water is almost equal to the one in air. Assume that the water thickness is 50 cm, What is the total energy lost by the shower electrons ($e^\pm$) crossing the detector? What is their mean pathlength in the detector?

Exercise 3.3.13

In 1974 a team led by S.C. Ting [6] carried out one of the experiments that demonstrated the existence of particles made of *charm* quarks. This experiment was run at the Brookhaven National Laboratory using 28 GeV protons hitting a beryllium target and studying the process

$$p + \text{Be} \rightarrow e^+ + e^- + X$$

where X indicates whatever particle or group of particles not detected. The invariant mass of the system (e^+, e^-), accessible through the identification of e^+ and e^- and the measurement of their momenta (both absolute value and direction), showed an evident peak at 3.1 GeV energy, later interpreted as the production and the consecutive decay of the J/ψ $(=c\bar{c})$ neutral meson having mass $m_J = 3.1$ GeV/c^2

$$J/\psi \rightarrow e^+ + e^-$$

Consider, for the sake of simplicity, that the observed process is

$$p + p \rightarrow J/\psi + p + p \tag{3.2}$$

assuming target protons to be at rest:

1. Calculate the energy threshold for the process (3.2).
2. As in the experiment, assume proton energy to be 28 GeV. Calculate the minimum and maximum production energy for the J/ψ.
3. Calculate the minimum opening angle of the e^+e^- pair.
4. One of the observed events has an electron with 10 GeV/c momentum and a positron at $\Delta\theta = 16°$. What is the positron momentum such that this pair can originate from the decay of the J/ψ?

Exercise 3.3.14

In an e^+e^- collider an experiment is carried out to study the tau-lepton production ($m_\tau = 1777$ MeV/c^2) at 29 GeV.

$$e^+ + e^- \rightarrow \tau^+ + \tau^-$$

(a) What is the energy of the τ's?
(b) Estimate the τ^+ mean lifetime, taking into account that the μ^+ has a mean lifetime of 2.2 μs, decays with 100% branching ratio (BR) as $\mu^+ \rightarrow e^+ + \nu_e + \bar{\nu}_\mu$ and the BR of the $\tau^+ \rightarrow e^+ + \nu_e + \bar{\nu}_\tau$ decay mode is about 18 %.
(c) The particle detection is performed in a cylindrical detector oriented along the two colliding beams and capable of tracking all charged particles. The detector has an internal radius of 5 cm to host the beam pipe. Is it possible to observe tau decays in this detector?

Exercise 3.3.15

An electromagnetic calorimeter is calibrated using different particle beams. The calorimeter is a sandwich of lead (Pb) and plastic scintillators. It is made up of consecutive stacks of one scintillator and one Pb slab, both 1 cm thick. Beams of 5 GeV electrons, muons and photons are used for calibration.

- Determine the mean energy deposited by electrons and muons in the fourth scintillator.
- Compare the results obtained using electron and photon beams. Explain how to discriminate between these two cases.

[Pb: $X_0 = 0.56$ cm, $\rho = 11$ g/cm^3 - Scintillator: $X_0 = 42$ cm, $\rho = 1.03$ g/cm^3]

Exercise 3.3.16

In the years 70's, a series of experiments were conducted to detect neutrinos and antineutrinos (in large mass neutrino detectors) emerging from thick targets. These experiments were called *beam dump* experiments [7].

Assume a copper target ($A = 63.5$, $\rho = 8.96$ g/cm^3), where an intense 400 GeV proton beam is dumped on. Emerging neutrinos can be interpreted as originating from the associate production of $c\bar{c}$ pairs, their fragmentation into charmed hadrons and their subsequent decays via leptonic or semi-leptonic channels (i.e. including neutrinos).

This technique is based upon the fact that all hadrons, produced in the dump, are absorbed but charmed particles, which are able to decay before interacting with the matter.

(a) Assuming that the proton cross-section depends on the target mass number A as

$$\sigma_{pA} = \sigma_{pp}\, A^{\frac{2}{3}}, \tag{3.3}$$

with $\sigma_{pp} \simeq 40$ mb, calculate the p-Cu interaction length.

(b) Assume for simplicity that all associated charm production processes can be represented as

$$p + p \rightarrow D^+ + D^- + X, \tag{3.4}$$

where X stands for any other particles involved in the reaction (3.4). Draw the flavor flow diagrams corresponding to the simplest choice of X.

(c) Which are the simplest leptonic and semi-leptonic decays of $D's$ giving rise to neutrinos? Are either neutrinos, ν_μ or ν_e, or antineutrinos, $\bar{\nu}_\mu$ or $\bar{\nu}_e$, produced from D^+? Draw the Feynman graphs for a decay into ν_μ and another into $\bar{\nu}_e$.

(d) Neutrinos are produced in the beam dump provided that $\lambda_{dec} \ll \lambda_{int}$, where λ_{dec} is the mean decay pathlength and λ_{int} is the $D^\pm$ interaction length in Copper. Evaluate if, and for which $D^\pm$ momenta, this condition is fulfilled in the reaction (3.4). For this purpose, assume that $D^\pm$ interaction length follows Eq. (3.3), with $\sigma_{Dp} \approx 30$ mb replacing σ_{pp}. Use also the following values $m_{D^\pm} = 1.87\,\text{GeV}/c^2$, $\tau(D^\pm) = 1.04 \times 10^{-12}$ s, $BR(D^\pm \rightarrow \nu_\mu) = BR(D^\pm \rightarrow \nu_e) = 17\%$.

(e) Which is the expectation for the ratio $\frac{\nu_\mu + \bar{\nu}_\mu}{\nu_e + \bar{\nu}_e}$?

Exercise 3.3.17

A typical experiment measures neutral pions through the detection of the photons emitted in the decay $\pi^0 \rightarrow \gamma\gamma$. Assume that the pion energies are around 1 GeV and that a 1 cm Pb plate is used for their conversion to electrons ($+$ and $-$), which are measured in a downstream detector. The Pb radiation length is 5.6 mm. What is the pion detection efficiency?

Hint:The detection efficiency is the probability that both photons from π^0-decays are converted in one or more electrons.

Exercise 3.3.18

The search for neutrino oscillations [8] is one of the hottest topics in recent years. Assume that the oscillation probability between electron and muon neutrino (or antineutrinos) is

$$P(\nu_e \rightarrow \nu_\mu) \simeq 0.20\ \sin^2\left(10^{-3}\ \frac{L[\text{m}]}{E[\text{MeV}]}\right) \tag{3.5}$$

where L is the distance between the neutrino production and detection points and E is the neutrino energy.

We aim to study the oscillation phenomenon operating a neutrino detector near a nuclear reactor. The nuclear reactor emits electron antineutrinos through the β^- decays of the radionuclides present in the reactor core at a rate of 10^{18} s^{-1}. The detector has a mass of 1 ton and is located at 200 m from the reactor core (both the reactor core and the detector can be considered point-like). Assume that the average energy of the antineutrinos from the reactor is 2 MeV and their detection efficiency is 70%. The total electron antineutrino cross section at 2 MeV is about 2 10^{-43} cm^2.

(a) Which are the possible reactions and the particles that is possible to detect for the two species of antineutrinos ($\bar{\nu}_e$, $\bar{\nu}_\mu$)? which are the interactions involved?
(b) If antineutrinos do not oscillate, how many interactions per year are measured?
(c) If they oscillate following Eq. (3.5), how many interactions per year are measured?
(d) In the latter case, what is the probability that the oscillation phenomenon is not observed (null result) in one year?

Exercise 3.3.19

A sequential experimental apparatus is used to analyze a charged particle. In the first part the particle trajectory in a magnetic field $B = 1$ Tesla is found to have a curvature corresponding to a sagitta $s_1 = 3$ cm measured along a track length $l_1 = 80$ cm. After passing though a passive medium the same particle is found to have a curvature of radius $R_2 = 121$ cm in the same field B. In the same region, a time-of-flight system performs a speed measurement which gives $v_2 = 2.8 \times 10^8$ m/s.

(a) Find the rest mass and kinetic energy of the particle before the slowing down.
(b) Evaluate the energy lost in the medium.
(c) The time-of-flight measurement is performed over a basis of 14 m. Repeating the same measurement with several equal particles, only 50% of the particles reach the last counter. Interpreting this loss as being due to the decay of this particle, compute its mean lifetime.

Exercise 3.3.20

A photon hits the wall of a liquid hydrogen bubble chamber (BC) producing an electron pair as shown in the figure below. The magnetic field $B = 0.8$ T is perpendicular to the plane of the figure, the density is $\rho = 0.071$ g/cm^3. The electron and positron tracks are detected in the BC as two opposite arcs whose measured diameters are 80 cm. The diameters are measured as the distance between the entrance and exit points for each particle.

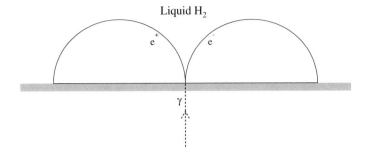

(a) Find the energy of the photon, neglecting the energy losses in the liquid hydrogen;
(b) estimate the same energy, taking into account these losses.

Hint: considering energy losses, at first approximation the track length is the same as in the case of no losses.

Exercise 3.3.21

Electron antineutrinos from nuclear reactors have typical energies E_ν of a few MeV (assume a continuous spectrum with a mean value of 2 MeV). They can be detected via the reaction $\bar{\nu}_e + p \rightarrow e^+ + n$ in a detector medium containing free protons. The process is observable because of the positron annihilation, $e^+ + e^- \rightarrow 2\gamma$ that follows the antineutrino interaction. Assume that the detector is large enough to measure the entire deposit of energy of the gammas. The detector medium (e.g. liquid scintillator) is surrounded by photomultipliers: the total energy, called *visible energy* E_{vis}, is measured and the neutrino energy is inferred.

(a) What is the dominating process for gamma energy deposit? Which is the characteristic length of the medium which determines if the detector is large enough?
(b) Estimate the kinetic energy of the recoiling neutron.
(c) Find the relation between E_ν and E_{vis}.
(d) Which is the minimum detectable neutrino energy?

References

1. Hagedorn, R.: Relativistic Kinematics. Literary Licensing, LLC (2012)
2. Tanabashi et al., M.: Particle data group. Phys. Rev. D **98**, 030001 (2018). http://pdg.lbl.gov/
3. Greisen, K.: End to the cosmic-ray spectrum? Phys. Rev. Lett. **16** 748 (1966); Zatsepin, G.T., Kuzmin, V.A.: Upper limit of the spectrum of cosmic rays. Sov. Phys. JETP Lett. **4**, 78 (1966)
4. Baroni, G., Di Liberto, S., Ginobbi, P., Petrera, S., Romano, G.: An attempt to detect particles of very short lifetimes produced in high-energy neutrino interactions. Lett. Nuovo Cim. **24**, 45 (1979)
5. Fukuda, S., et al.: The Super-Kamiokande detector. Nucl. Instr. Meth. A **501**, 418 (2003)
6. Aubert, J.J., et al.: Experimental observation of a heavy particle J. Phys. Rev. Lett. **33**, 1404 (1974)

7. Hansl, T., et al.: Results of a beam dump experiment at the CERN SPS neutrino facility. Phys. Lett. **74B**, 139 (1978)
8. Bettini, A.: Beyond the standard model. In: Introduction to Elementary Particle Physics, pp. 354–385. Cambridge University Press, Cambridge (2008)

Correction to: Problems and Solutions in Nuclear and Particle Physics

Correction to:
S. Petrera, *Problems and Solutions*
***in Nuclear and Particle Physics*,**
https://doi.org/10.1007/978-3-030-19773-5

The original version of the book frontmatter (Preface and Notes) were revised. The references lists had been placed at the end of Preface and Notes for clarity in understanding.

The original version of the frontmatter was revised: The correction is available at
https://doi.org/10.1007/978-3-030-19773-5

© Springer Nature Switzerland AG 2019   C1
S. Petrera, *Problems and Solutions in Nuclear and Particle Physics*,
UNITEXT for Physics, https://doi.org/10.1007/978-3-030-19773-5_4

Appendix
Solutions of Exercises and Problems

A.1 Solutions of Nuclear Physics (Chapter 1)

1.1 Initial Problems

Exercise 1.1.1

To give a rough estimate of the nuclear density, we assume that

– the binding energy is negligible;
– proton and neutron have the same mass, $m_p = m_n$;
– the nuclear radius is $R = r_0 \cdot A^{1/3}$, with $r_0 = 1.2$ fm.

Under these assumptions we have

$$\rho = \frac{M}{V} = \frac{A \cdot m_p}{4/3 \, \pi \, r_0^3 \, A} = \frac{3 \, m_p}{4 \, \pi \, r_0^3} \simeq \frac{3 \times 1.67 \; 10^{-24} \text{g}}{12.56 \, (1.2 \; 10^{-13} \text{cm})^3} \simeq 2.3 \cdot 10^{14} \, \text{g/cm}^3.$$

Exercise 1.1.2

The electrostatic energy for a charge Q distributed uniformly in a sphere of radius R is $3/5 \cdot Q^2/(4\pi\epsilon_0 R)$. Equating this energy to the Coulomb binding energy in the *SEMF* we get ,

$$\frac{3 \, Z^2 e^2}{20 \, \pi \epsilon_0 \, r_0 A^{1/3}} = a_C \cdot \frac{Z^2}{A^{1/3}},$$

and then

$$a_C = \frac{3}{5} \times \frac{e^2}{4\pi\epsilon_0} \times \frac{1}{r_0} = \frac{3}{5} \times \alpha \hbar c \times \frac{1}{r_0} \simeq 0.6 \times \frac{197 \text{ MeV fm}}{137} \times \frac{1}{1.2 \text{ fm}} \simeq 0.7 \text{ MeV}.$$

© Springer Nature Switzerland AG 2019
S. Petrera, *Problems and Solutions in Nuclear and Particle Physics*,
UNITEXT for Physics, https://doi.org/10.1007/978-3-030-19773-5

Exercise 1.1.3

Using the result of problem 1.1.1, we assume for the nuclear density $\rho \simeq 2.3 \times 10^{14}$ g/cm^3. Denoting with R and M respectively the radius and mass of the neutron star, from the relation

$$\frac{4}{3} \pi R^3 \rho = M \approx M_\odot$$

we obtain

$$R \approx \left(\frac{3M_\odot}{4\pi\rho}\right)^{1/3} \simeq \left(\frac{3 \cdot 2 \ 10^{33}\text{g}}{4 \cdot 3.14 \cdot 2.3 \ 10^{14} \text{ g/cm}^3}\right)^{1/3} \simeq 12.8 \text{ km.}$$

Exercise 1.1.4

Let us consider two deuterons moving along a certain direction with equal but opposite velocities (head-on collision). Since the motion is thermal, the kinetic energy of each deuteron can be treated as non-relativistic, $E = 1/2 M v^2$, and assumed to be of the order of $k_B T$.

At large distance, in the rest frame of one of the deuterons, the other has velocity $2v$. The corresponding kinetic energy equates the repulsive electrostatic energy at the minimum distance, because of energy conservation

$$\frac{1}{2} M (2v)^2 = 4E = 4 k_B T = \frac{1}{4\pi\epsilon_0} \frac{e^2}{r_{min}}.$$

Thus a rough estimate of the minimum temperature to get nuclear processes is

$$T_{min} = \frac{e^2}{4\pi\epsilon_0} \frac{1}{4 k_B r_{min}} = \frac{\alpha \hbar c}{4 k_B r_{min}} \simeq$$

$$\simeq \frac{197 \text{ MeV} \cdot \text{fm}}{137 \times 4 \times 8.6 \ 10^{-11}\text{MeV} \cdot \text{K}^{-1} \times 1 \text{ fm}} \simeq 4 \ 10^9 \text{ K.}$$

Exercise 1.1.5

The neutron rate per solid angle is

$$\frac{dN}{dt \, d\Omega} = \frac{d\sigma}{d\Omega} \frac{dn_b}{dt} n_T L$$

where $dn_b/dt = I/e$ is the deuteron beam intensity and n_T is the number of target nuclei per unit volume, $n_T = N_A/A \, \rho$. The solid angle between the detector and the interaction region (assumed point-like) is $\Delta\Omega = S/R^2$. Then we have

$$\frac{dN}{dt} = \frac{d\sigma}{d\Omega} \frac{S}{R^2} \frac{I}{e} \frac{N_A}{A} \rho L \simeq$$

$$13 \ 10^{-3} \ 10^{-24}\text{cm}^2/\text{sr} \ \frac{20}{300^2} \ \frac{2 \ 10^{-6} \text{ A}}{1.6 \ 10^{-19}\text{C}} \ \frac{6 \ 10^{23}}{3} \ 0.2 \ 10^{-3} \text{ g/cm}^2 \simeq 1.4 \ 10^3 \text{ s}^{-1}$$

1.2   Nuclear Scattering

Exercise 1.2.1

(1) The rate of electrons scattered in the solid angle $\Delta\Omega$ around the angle θ, from a beam of intensity dn_b/dt (e/s) incident perpendicularly on a target with atomic number A, thick x_T (g/cm^2), is

$$\frac{dn}{dt} = \frac{dn_\mathrm{b}}{dt} \times \frac{dn_\mathrm{T}}{dS} \times \int_{\Delta\Omega} \frac{d\sigma}{d\Omega}\, d\Omega \simeq \frac{I_e}{e} \times x_\mathrm{T} \frac{N_A}{A} \times \frac{S}{R^2} \times \frac{d\sigma}{d\Omega}(\theta),$$

being $\sqrt{S}/R \ll 1$. Then we have

$$\frac{dn}{dt} \simeq \frac{5\ 10^{-6}}{1.6\ 10^{-19}} \times \frac{0.12 \cdot 6.02\ 10^{23}}{40} \times \frac{20}{100^2} \times \frac{d\sigma}{d\Omega}(\theta) \simeq$$

$$\simeq 1.13\ 10^{32}\ \frac{\mathrm{sr}}{\mathrm{cm}^2 \cdot \mathrm{s}} \times \frac{d\sigma}{d\Omega}(\theta).$$

$d\sigma/d\Omega\,(\theta)$ is given by $|F(q^2)|^2 \times (d\sigma/d\Omega)_\mathrm{Mott}$. For $\beta \to 1$ the Mott cross section at 40° is

$$\left(\frac{d\sigma}{d\Omega}\right)_\mathrm{Mott} = \frac{Z^2 \alpha^2 (\hbar c)^2 \cos^2 \theta/2}{4\,(p\,c)^2\ \sin^4 \theta/2} \simeq \left(\frac{20 \times 197}{137}\right)^2$$

$$\times \frac{\cos^2 20°}{4 \times 700^2 \times \sin^4 20°}\ \mathrm{fm}^2/\mathrm{sr} \simeq 0.272\ \mathrm{mb/sr}$$

The form factor for a uniform charge distribution in a sphere of radius R_A is

$$F(q^2) = 3\,\frac{\sin x - x \cos x}{x^3},$$

where $x = q\,R_A/\hbar$

$$x = \frac{2\,p\,c\ \sin\theta/2 \times (1.18\,A^{1/3} - 0.48)\ \mathrm{fm}}{\hbar\,c} \simeq$$

$$\simeq \frac{2 \times 700\ \mathrm{MeV} \times \sin 20° \times 3.56\ \mathrm{fm}}{197\ \mathrm{MeV\ fm}} \simeq 8.64,$$

Hence we get $F(q^2) \simeq 3.18\ 10^{-2}$ and finally obtain

$$\frac{dn}{dt} \simeq 1.13\ 10^{32}\ \frac{\mathrm{sr}}{\mathrm{cm}^2 \cdot \mathrm{s}} \times 0.272\ 10^{-27} \frac{\mathrm{cm}^2}{\mathrm{sr}} \times (3.18\ 10^{-2})^2 \simeq 31\ \mathrm{electrons/s}.$$

In Fig. 1.1 the rate of the scattered electrons is shown as a function of the angle.

Fig. 1.1 Rate
(counts-per-sec) for 700
MeV/c electron scattering
against ^{40}C

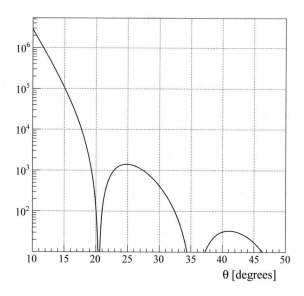

(2) As it can be seen in the figure the first local maximum is at about 25°. Here the detector delivers about 1400 counts per second. The mean number of electron-ion pairs produced by an electron crossing the gas mixture is

$$N_e = \frac{(-dE/dx)_{ion} \times \rho \times d}{W_{ion}} \times \epsilon_{ion}$$

$$\simeq \frac{1.4 \times 2 \ 10^6 \ \text{eV}/(\text{g cm}^{-2}) \times 1.8 \ 10^{-3}\text{g/cm}^3 \times 0.1 \ \text{cm}}{15 \ \text{eV}} \times 0.10 \simeq 3.36,$$

where we used 2 MeV/(g cm^{-2}) for the *minimum ionization* energy loss.

The number of events for which *no electron* reaches the anode is

$$\epsilon_0 = (1 - P)^{N_e} \simeq 0.70^{3.36} \simeq 30.2\%.$$

The rate of coincident counts is finally

$$\frac{dn_c}{dt} = \frac{dn}{dt} \times (1 - \epsilon_0)^2 \simeq 1400 \times 0.698^2 \simeq 1400 \times 0.49 \simeq 690 \ \text{counts/s.}$$

Exercise 1.2.2

The number of minima is given by number of the zeroes of the form factor for a uniform charge distribution. The latter is given by

$$F(q^2) = 3 \frac{\sin x - x \cos x}{x^3},$$

Fig. 1.2 tan x versus x (black). $y = x$ (red)

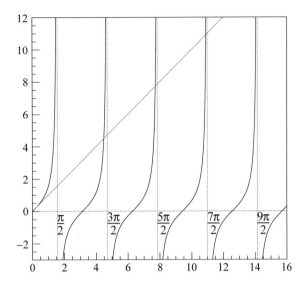

where $x = q\,R/\hbar$, with R given by the radius of the nucleus, $R = (1.18 A^{1/3} - 0.48)$ fm. $F(q^2) = 0$ leads to the equation

$$\tan x = x.$$

A graphical method allows to estimate the positions of the zeroes (see Fig. 1.2, black: $\tan x$, red: x) as the ones where the tangent equates the straight line. This occurs close to $x \simeq 3\pi/2, 5\pi/2, 7\pi/2, 9\pi/2 \ldots$

In the actual experimental conditions x is limited up to a maximum $x_{max} = q_{max} R/\hbar$. Remembering that

$$q = 2\,p \cdot \sin \frac{\theta}{2} \quad \Longrightarrow \quad q_{max} = 2\,p \simeq 2\,\frac{E}{c}$$

we have

$$x_{max} = 2\,\frac{E}{\hbar c} \times (1.18\,A^{1/3} - 0.48)\ \text{fm} \simeq \frac{2 \cdot 180\ \text{MeV}}{197\ \text{MeV fm}} \times 6.4\ \text{fm} \simeq 11.7.$$

There are three minima below this value, corresponding to the zeroes up to $7\pi/2$.

Exercise 1.2.3

Considering the Rutherford cross section, we can write the counting rate at angle θ as

$$f(\theta) = K \frac{\Phi}{\sin^4 \theta/2},$$

where Φ is the incident flux and K an overall factor including various terms (kinematical, geometrical, etc.). We assume that

$$f(20°) = K \frac{\Phi}{\sin^4(20°/2)} = 1 \text{ s}^{-1}. \tag{1.1}$$

Denoting with f_a the counting rate for a flux attenuated by a factor a, we have

$$f_a(10°) = K \frac{a \, \Phi}{\sin^4(10°/2)} = 1 \text{ s}^{-1}, \tag{1.2}$$

Dividing (1.2) by (1.1) we get

$$a = \left(\frac{\sin 5°}{\sin 10°} \right)^4 \simeq 6.3\%.$$

Using the attenuated beam, the counting rate at 20° is $f_a(20°)$. The mean waiting time is its inverse

$$\langle \Delta t \rangle = \frac{1}{f_a(20°)} = \frac{1}{a \, f(20°)} = \frac{1}{0.063 \times 1 \text{ s}^{-1}} \simeq 16 \text{ s}.$$

Exercise 1.2.4

The differential cross section $d\sigma/d\Omega(\theta)$ is given by $|F(q^2)|^2 \times (d\sigma/d\Omega)_{\text{Mott}}$. For $\beta \to 1$ the latter cross section is

$$\left(\frac{d\sigma}{d\Omega} \right)_{\text{Mott}} = \frac{Z^2 \alpha^2 (\hbar c)^2 \cos^2 \theta/2}{4 (p \, c)^2 \, \sin^4 \theta/2}$$

$$\simeq \left(\frac{6 \times 197}{137} \right)^2 \times \frac{\cos^2 7.5°}{4 \times 100^2 \times \sin^4 7.5°} \text{ fm}^2/\text{sr} \simeq 6.3 \times 10^{-26} \text{ cm}^2/\text{sr}.$$

The form factor can be neglected because the momentum transfer is small.[1] Since $\sqrt{S}/R \ll 1$, we can simply write

$$\sigma = \left(\frac{d\sigma}{d\Omega} \right)_{\text{Mott}} \times S/R^2 \simeq 6.3 \cdot 10^{-26} \text{ cm}^2/\text{sr} \times 7.5 \cdot 10^{-4} \text{sr} \simeq 47 \text{ μb}.$$

---

[1]The relevant argument for the form factor is $x = q \, R_A/\hbar$, where $R_A \simeq 1.2 \text{ fm} \times A^{1/3}$ is the nuclear radius. Hence $x \simeq 0.36$ and for a uniform charge distribution, we have

$$F(q^2) = 3 \frac{\sin x - x \cos x}{x^3} \simeq 0.99.$$

The number of scatterers per unit surface is $dn_T/dS = d \times N_A/A \simeq 5 \times 10^{22}$ cm^{-2}. Then we have

$$\frac{dN_e}{dt} = \frac{I_0}{e}\sigma\frac{dn_T}{dS} \simeq 1.5 \times 10^8 \text{ s}^{-1}.$$

Exercise 1.2.5

The momentum transfer (we have $pc \simeq E$) is

$$q = 2\,p \cdot \sin\theta/2 = 2 \times 500 \times \sin 5° \simeq 87.2 \text{ MeV/c}.$$

The Mott cross section at $10°$ can be written as

$$\left(\frac{d\sigma}{d\Omega}\right)_{\text{Mott}} = 4\,\frac{Z^2\alpha^2(\hbar c)^2}{(q\,c)^4}E^2\cos^2\frac{\theta}{2} \simeq$$

$$\simeq 4 \times \left(\frac{26 \times 197}{137}\right)^2 \times \frac{500^2 \cos^2 5°}{87.2^4} \text{ fm}^2/\text{sr} \simeq 0.24 \text{ b/sr}$$

The form factor is given by

$$F(q^2) = 3\,\frac{\sin x - x\cos x}{x^3},$$

where $x = q\,R_A/\hbar$. Using $R_A \simeq 1.2$ fm $\times A^{1/3}$ for the nucleus radius, we find $x = 87.2 \times 4.6 / 197 \simeq 2$ and then

$$\frac{d\sigma}{d\Omega} = \left(\frac{d\sigma}{d\Omega}\right)_{\text{Mott}}|F(q^2)|^2 = 0.24\,\frac{\text{b}}{\text{sr}} \times 0.65^2 \simeq 0.10\,\frac{\text{b}}{\text{sr}}$$

Exercise 1.2.6

The Rutherford cross section can be written as

$$\frac{d\sigma}{d\Omega} = \left[\frac{zZ\alpha(\hbar c)}{4E_\alpha}\right]^2\frac{1}{\sin^4\frac{\theta}{2}},$$

where z and E_α are respectively the charge and kinetic energy of alpha particles. The solid angle corresponding to the detector is

$$\Delta\Omega = \frac{S}{R^2} = 10^{-3} \text{ sr}.$$

To achieve the required accuracy, we calculate the cross section at the largest angle ($150°$) in the chosen interval, where it has the smallest value

$$\sigma = \left[\frac{zZ\alpha(\hbar c)}{4E_\alpha}\right]^2\frac{\Delta\Omega}{\sin^4\frac{\theta}{2}} \simeq \left[\frac{2 \times 79 \times 197}{137 \times 4 \times 5.5}\right]^2\frac{10^{-3}}{0.8705} \simeq 0.12 \text{ fm}^2 = 1.2 \times 10^{-27} \text{ cm}^2.$$

The event rate in the detector is

$$r = I_\alpha \rho \Delta l \frac{N_A}{A} \sigma$$

where I_α is the beam intensity ($N_A = 6.02 \times 10^{23}$ mole^{-1} is the Avogadro number). Thus the intensity of the α beam must be

$$I_\alpha > \frac{A}{\rho \Delta l N_A} \frac{r}{\sigma} = \frac{197}{0.1 \times 6.02 \times 10^{23}} \frac{10}{1.2 \times 10^{-27} \text{ cm}^2} \simeq 2.7 \times 10^7 \text{ s}^{-1}.$$

Exercise 1.2.7

The Q-factor of the reaction $p + {}_3^7\text{Li} \rightarrow {}_2^4\text{He} + {}_2^4\text{He}$ is

$$Q = M_p + M_{\text{Li}} - 2M_\alpha = M_p + [3M_p + 4M_n - B({}_3^7\text{Li})] - 2[2M_p + 2M_n - B(\alpha)] =$$

$$= 2B(\alpha) - B({}_3^7\text{Li}) = 2 \cdot 28.3 - 39.3 = 17.3 \text{ MeV} > 0$$

The reaction is exothermic.

According to the shell model, the ${}_3^7\text{Li}$ shell occupancies for protons and neutrons are

$$p : (1s^{1/2})^2 (1p^{3/2})^1$$

$$n : (1s^{1/2})^2 (1p^{3/2})^2$$

The spin-parity is then determined by the odd $(1p^{3/2})$ proton shell and is $J^P = (3/2)^-$.

Protons at rest cannot interact with ${}_3^7\text{Li}$ nuclei because of the Coulomb barrier. Neglecting for simplicity the size of the proton with respect to the one of ${}_3^7\text{Li}$, the minimum proton kinetic energy turns out to be

$$T_{\text{min}} = \frac{zZe^2}{4\pi\epsilon_0 d} \simeq \frac{zZ\alpha\hbar c}{R({}_3^7\text{Li})} \simeq \frac{3 \cdot 197 \text{ MeV fm}}{137 \cdot 1.2 \, 7^{1/3} \text{ fm}} \simeq 1.9 \text{ MeV}.$$

Indicating the spin-parities of the nuclei involved in the reaction, we have

$$p\left(\frac{1}{2}^+\right) + {}_3^7\text{Li}\left(\frac{3}{2}^-\right) \rightarrow {}_2^4\text{He}(0^+) + {}_2^4\text{He}(0^+)$$

Knowing that the final orbital angular momentum is zero, we deduce that the initial total angular momentum must be zero. The angular momentum conservation imposes

$$\frac{1}{2} \oplus \frac{3}{2} \oplus L_i = 0$$

denoting with $\oplus$ the operation of addition of angular momenta and with L_i the initial orbital angular momentum. Since $\frac{1}{2} \oplus \frac{3}{2} = 1, 2$, then it follows that L_i must be either 1 or 2.

On the other hand parity conservation imposes the same parity for the initial and final states. The final parity is evidently $+1$ and then

$$P_i = P(p) \times P(^7_3\text{Li}) \times P_{\text{orb}} = (+1) \times (-1) \times (-1)^{L_i}$$

hence L_i must be odd and finally $L_i = 1$.

Exercise 1.2.8

The Mott cross section at $5°$ is

$$\left(\frac{d\sigma}{d\Omega}\right)_{\text{Mott}} = \left(\frac{Z\alpha \hbar c}{p\,c}\right)^2 \frac{\cos^2 \theta/2}{4 \sin^4 \theta/2} \simeq$$

$$\simeq \left(\frac{6 \times 197}{137 \times 720}\right)^2 \times \frac{\cos^2 2.5°}{4 \times \sin^4 2.5°} \frac{\text{fm}^2}{\text{sr}} \simeq 99 \frac{\text{mb}}{\text{sr}}$$

Using the measured cross section, we derive the absolute square of the form factor as

$$|F(q^2)|^2 = \frac{\left(\frac{d\sigma}{d\Omega}\right)_{\text{meas}}}{\left(\frac{d\sigma}{d\Omega}\right)_{\text{Mott}}} \simeq \frac{80}{99} \simeq 0.808$$

If the nucleus is spherically symmetric, the form factor is real and then it can be obtained as the square root of its absolute square. Note that the indetermination of the sign is resolved looking at the momentum transfer. At $5°$ it is small and the form factor is still far from the first zero. Thus we can assume that the form factor is positive.

The momentum transfer turns out to be

$$q = 2 p \sin\frac{\theta}{2} \simeq 62.8 \text{ MeV/c}$$

and then

$$\langle r^2 \rangle = \frac{6\hbar^2}{q^2} [1 - F(q^2)] \simeq \frac{6 \ 197^2}{62.8^2} \times (1 - \sqrt{0.808}) \simeq 5.95 \text{ fm}^2$$

from which we get for the ^{12}C nuclear radius 2.44 fm.

Exercise 1.2.9

For $\beta \simeq 0.1$ a proton is non-relativistic and its maximum energy can be written as

$$T_{\text{max}} \simeq \frac{1}{2} m_p \beta^2 \simeq 0.5 \times 938 \times 0.01 \simeq 4.7 \text{ MeV.} \tag{1.3}$$

The maximum value of beta (and then of energy) corresponds to the forward scattering with the incident particle (γ or n) scattered back. In the case of photon scattering, $\gamma + p \rightarrow \gamma + p$, the kinematic relation is the same as in the Compton scattering, with the proton replacing the electron.

$$E'_\gamma = \frac{E_\gamma}{1 + E_\gamma/m_p(1 - \cos\theta)},$$

where E_γ and E'_γ are the photon initial and final energies and θ is the photon scattering angle. For $\theta = 180°$ the proton energy is maximum. Denoting with $T = E_\gamma - E'_\gamma$ the proton kinetic energy, its maximum is

$$T_{max} = E_\gamma - \frac{E_\gamma}{1 + 2E_\gamma/m_p} = \frac{2E_\gamma^2}{m_p + 2E_\gamma} \tag{1.4}$$

(a) If we assume 10 MeV for the photon energy, from (1.4) we have

$$T_{max} = \frac{2E_\gamma^2}{m_p + 2E_\gamma} \simeq \frac{2 \times 100}{938 + 20} \simeq 0.21 \text{ MeV},$$

which is largely smaller than (1.3).
(b) Solving Eq. (1.4) in E_γ and assuming that the proton energy is the one measured (1.3), the photon energy is

$$E_\gamma = \frac{T_{max} + \sqrt{T_{max}(T_{max} + 2m_p)}}{2} \simeq \frac{4.7 + \sqrt{4.7(4.7 + 2 \cdot 938)}}{2} \simeq 49 \text{ MeV}.$$

(c) In the case of neutron scattering, $n + p \rightarrow n + p$, the maximum proton energy occurs again in the forward scattering. Since the proton and neutron masses are very similar, this corresponds to the neutron stopped and the proton achieving the entire initial energy. This is a well known fact in a billiard, but can be easily obtained from kinematics. Being $\theta = 0$, we can treat the kinematics in one dimension as

$$T_n = T'_n + T'_p \qquad p_n = p'_n + p'_p.$$

Expressing the kinetic energies as $T = p^2/(2m)$ and assuming $m_n = m_p$, we get

$$p'_p(p'_p - p_n) = 0 \qquad \Longrightarrow \qquad p'_p = p_n \simeq 4.7 \text{ MeV}.$$

Exercise 1.2.10

The Rutherford cross section is

$$\frac{d\sigma}{d\Omega} = \left[\frac{zZ\alpha(\hbar c)}{4E_\alpha}\right]^2 \frac{1}{\sin^4\frac{\theta}{2}},$$

where z and E_α are respectively the charge and kinetic energy of alpha particles. Using the solid angle subtended by the detector

$$\Delta\Omega = \frac{S}{R^2} = \frac{0.5}{10^2} = 5 \ 10^{-3} \text{ sr},$$

we get for the cross section at angle θ

$$\sigma(\theta) = \left[\frac{zZ\alpha(\hbar c)}{4E_\alpha}\right]^2 \frac{\Delta\Omega}{\sin^4 \frac{\theta}{2}} \simeq \left[\frac{2 \times 79 \times 197}{137 \times 4 \times 5.64}\right]^2 \frac{5 \ 10^{-3}}{\sin^4 \frac{\theta}{2}} \simeq$$

$$\simeq \frac{0.507}{\sin^4 \frac{\theta}{2}} \text{ fm}^2 = \frac{5.07 \times 10^{-27}}{\sin^4 \frac{\theta}{2}} \text{ cm}^2.$$

The number of counts during the time interval Δt is

$$N(\theta) = I_\alpha \Delta t \rho d \frac{N_A}{A} \sigma(\theta)$$

where I_α is the α beam intensity and d is the target thickness. Solving in I_α we have

$$I_\alpha = \frac{N(\theta)}{d\rho\Delta t \frac{N_A}{A}\sigma(\theta)} = \frac{197}{0.005 \times 19.3 \times 3600 \times 6.02 \ 10^{23} \times 5.07 \ 10^{-27}} \times N(\theta)\sin^4 \frac{\theta}{2}$$

and then

$$I_\alpha \simeq N(\theta) \times 186 \times \sin^4 \frac{\theta}{2} \text{ s}^{-1}.$$

The following table gives the beam intensity in s^{-1} resulting at each angle

θ	15°	25°	35°	45°	55°	65°	75°
I_α (s^{-1})	229.9 ± 3.5	242.1 ± 9.9	226 ± 19	199 ± 28	262 ± 47	201 ± 56	179 ± 67

The statistical errors are calculated as $\sqrt{N(\theta)} \times 186 \times \sin^4 \theta/2$. Calculating the weighted average and its variance we get

$$I_\alpha = 230.7 \pm 3.2 \text{ s}^{-1}.$$

1.3   Nuclear Binding Energy

Exercise 1.3.1

Decays among isobar nuclei belong to the class of beta decays. In the present case the mass number, $A = 197$, is odd so that there is only one stable nucleus. In fact, using the *semi-empirical mass formula (SEMF)* the atomic mass $M(A, Z)$ as a function of Z is a single curve, because the *pairing* term is null for all isobars. The stable nucleus has $Z_s = 79$, the nucleus with $Z = Z_s - 1 = 78$ can transmute to it via β^- decay whereas the nucleus with $Z = Z_s + 1 = 80$ can do it via β^+ decay or electron capture (EC).

We can write the atomic mass of $A = 197$ nuclei as

$$M(197, Z) = Zm_p + (197 - Z)m_n - B(197, Z)/c^2 + Zm_e,$$

where $B(197, Z)$ is the nuclear binding energy, for which we use the *SEMF*. Writing explicitly only the terms depending on Z we have

$$M(197, Z)\, c^2 = \text{const} + Z\,(m_p - m_n + m_e)\, c^2 + a_C\, \frac{Z^2}{197^{1/3}} + a_A\, \frac{(197 - 2Z)^2}{197} \simeq$$

$$\simeq \text{const} - 0.782\, Z + 0.697\, \frac{Z^2}{5.82} + 23.3\, \frac{(197 - 2Z)^2}{197}\ \text{MeV}.$$

For the β^- transition from $^{197}_{78}\text{Pt}$ we have $M(197, 78) - M(197, 79) \simeq 0.90$ MeV, hence it is allowed.

The β^+ transition from $^{197}_{80}\text{Hg}$ is allowed if $M(197, 80) - M(197, 79) > 2m_e$, if instead this difference is positive only the electron capture is possible. In the present case we have $M(197, 80) - M(197, 79) \simeq 0.3$ MeV/c^2.

As a conclusion the possible decay types are

$$\beta^-:\quad ^{197}_{78}\text{Pt} \rightarrow\ ^{197}_{79}\text{Au} + e^- + \bar{\nu}_e$$

$$EC:\quad e^- +\ ^{197}_{80}\text{Hg} \rightarrow\ ^{197}_{79}\text{Au} + \nu_e$$

Exercise 1.3.2

The mean neutron kinetic energy is $\langle E \rangle \approx k_B T \simeq k_B\, 300 \simeq 25$ meV. From the *semi-empirical mass formula* we get

$$B(235, 92) = 1786.8\ \text{MeV}$$

$$B(148, 57) = 1209.8\ \text{MeV}$$

$$B(87, 35) = 745.4\ \text{MeV}.$$

The neutron energy is negligible with respect to the other energies and then we have for the energy release

$$Q = B(148, 57) + B(87, 35) - B(235, 92) \simeq 168 \text{ MeV}.$$

A similar though less accurate conclusion can be reached using the B/A values from the binding energy per nucleon plot reported in all the textbooks. The values are 7.6, 8.2 and 8.6 MeV, respectively $A = 235$, 148 e 87. Hence we obtain

$$Q = 148 \cdot 8.2 + 87 \cdot 8.6 - 235 \cdot 7.6 \simeq 176 \text{ MeV}.$$

Exercise 1.3.3

At large distance, in the rest frame of one of the nuclei, the other has velocity $2v$. At the minimum distance R the two nuclei are at rest. From energy conservation then we get

$$\frac{1}{2}M(2v)^2 = 4E = \frac{e^2}{4\pi\epsilon_0 R} = \frac{\alpha\hbar c}{R} = \frac{197 \text{ MeV fm}}{137 \cdot 1.4 \text{ fm}} \simeq 1 \text{ MeV},$$

E, the mean kinetic energy of each nucleus, is then about 0.25 MeV.

Knowing that for $T = 300$ K the mean kinetic energy is $k_B T \simeq 25$ meV, the temperature for which nuclei have $E \simeq 0.25$ MeV is

$$T = \frac{E}{k_B} = \frac{0.25 \; 10^6 \text{ eV} \times 300}{25 \; 10^{-3} \text{ eV}} \simeq 3 \; 10^9 \text{ K}.$$

The energy release is

$$Q = B_T - 2B_D \simeq 4 \text{ MeV}.$$

Exercise 1.3.4

The reaction in the text belongs to the more general class

$$\nu_e + (A, Z) \rightarrow (A, Z+1) + e^-.$$

The threshold energy is given by

$$E_{\text{th}} = \frac{(m_e + M')^2 - M^2}{2M}, \tag{1.5}$$

where M and M' are the masses of (A, Z) and $(A, Z + 1)$ nuclei respectively. These masses are related to the binding energies as follows

$$M = ZM_p + (A - Z)M_n - B(A, Z)/c^2$$

$$M' = (Z+1)M_p + (A - Z - 1)M_n - B(A, Z+1)/c^2 = M + \Delta M,$$

hence we have

$$\Delta M = (M_p - M_n) + \Delta B/c^2 \quad \text{with} \quad \Delta B = B(A, Z) - B(A, Z+1).$$

ΔB can be calculated using the *semi-empirical mass formula*. In particular for odd-A nuclei only the Coulomb and asymmetry terms are needed, because:

1. the volume and surface terms depend only on A and they cancel out in the difference;
2. for odd-A the pairing term is null for both initial and final nuclei.

We have

$$\Delta B = -a_C \left\{ \frac{Z^2}{A^{1/3}} - \frac{(Z+1)^2}{A^{1/3}} \right\} - a_A \left\{ \frac{(A-2Z)^2}{A} - \frac{[A - 2(Z+1)]^2}{A} \right\} =$$

$$= a_C \frac{2Z+1}{A^{1/3}} - 4a_A \frac{A - 2Z - 1}{A}$$

In the reaction considered in the text $A = 37$ and $Z = 17$ and then we have

$$\Delta M = -1.293 + 0.697 \times \frac{35}{37^{1/3}} - 4 \times 23.3 \times \frac{2}{37} \simeq 1 \text{ MeV}.$$

Substituting this value in (1.5) we obtain

$$E_{\text{th}} = \frac{[m_e + (M + \Delta M)]^2 - M^2}{2M} = \frac{m_e(m_e + 2M) + \Delta M(2m_e + \Delta M + 2M)}{2M}$$

$$\simeq m_e + \Delta M \simeq 1.5 \text{ MeV}.$$

Exercise 1.3.5

Denoting by Q_- the Q-factor for the β_- decay

$$^{64}_{29}\text{Cu} \rightarrow \, ^{64}_{30}\text{Zn} + e^- + \bar{\nu}_e$$

and Q_+ the one for the β_+ decay

$$^{64}_{29}\text{Cu} \rightarrow \, ^{64}_{28}\text{Ni} + e^+ + \nu_e$$

we have (omitting the factor c^2 in the mass terms)

$$Q_- = 29M_p + 35M_n - B(64, 29) - 30M_p - 34M_n + B(64, 30) - m_e$$

$$= M_n - M_p - m_e + B(64, 30) - B(64, 29) \simeq 0.782 \text{ MeV} + B(64, 30) - B(64, 29).$$

Similarly we have

$$Q_+ = M_p - M_n - m_e + B(64, 28) - B(64, 29) \simeq -1.804 \text{ MeV} + B(64, 28) - B(64, 29).$$

From the *SEMF* we get

$$B(64, 30) - B(64, 29) = -0.697 \times \frac{30^2 - 29^2}{64^{1/3}} - 23.3 \times \frac{(64 - 60)^2 - (64 - 58)^2}{64} +$$

$$+ \frac{12 + 12}{\sqrt{64}} \simeq 0.0005 \text{ MeV},$$

$$B(64, 28) - B(64, 29) = -0.697 \times \frac{28^2 - 29^2}{64^{1/3}} - 23.3 \times \frac{(64 - 56)^2 - (64 - 58)^2}{64} +$$

$$+ \frac{12 + 12}{\sqrt{64}} \simeq 2.74 \text{ MeV}.$$

Hence we have

$$Q_- \simeq 0.78 \text{ MeV} \qquad Q_+ \simeq 0.94 \text{ MeV}.$$

Both decays are allowed. The maximum kinetic energies of the electron and positron are equal respectively to Q_- and Q_+.

Exercise 1.3.6

The stability condition can be written as $\frac{\partial M(A, Z)}{\partial Z} = 0$, $M(A, Z)$ being the atomic mass of the nucleus (A, Z), which is is single function for odd-A nuclei. Using the *SEMF* we have

$$\frac{2a_c Z}{A^{1/3}} - \frac{4a_a(A - 2Z)}{A} - (M_n - M_p - m_e)c^2 = 0$$

hence we get for the asymmetry coefficient

$$a_A = \frac{A}{4(A - 2Z)} \left[\frac{2a_c Z}{A^{1/3}} - (M_n - M_p - m_e)c^2 \right] \simeq 24 \text{ MeV}.$$

Exercise 1.3.7

For a β^+ decay, $(A, Z) \rightarrow (A, Z - 1) + e^+ + \nu_e$, the Q_β value, is

$$Q_\beta = [M(A, Z) - M(A, Z - 1) - m]c^2,$$

where

$$M(A, Z) = ZM_p + (A - Z)M_n - B(A, Z)/c^2$$

$$M(A, Z - 1) = (Z - 1)M_p + (A - Z + 1)M_n - B(A, Z - 1)/c^2.$$

Hence we have

$$Q_\beta = [M_p - M_n - m]c^2 - \Delta B, \tag{1.6}$$

where

$$\Delta B = B(A, Z) - B(A, Z - 1).$$

Calculating ΔB from the *SEMF*, we observe that all terms cancel but the Coulomb and asymmetry ones because

1. the volume and surface terms depend on A only,
2. A is odd and the pairing term is the same ($=0$) for both nuclei.

Then we have

$$\Delta B = -a_C \left\{ \frac{Z^2}{A^{1/3}} - \frac{(Z-1)^2}{A^{1/3}} \right\} - a_A \left\{ \frac{(A-2Z)^2}{A} - \frac{[A-2(Z-1)]^2}{A} \right\}$$

$$= -a_C \frac{2Z-1}{A^{1/3}} - 4a_A \frac{A-2Z+1}{A} \tag{1.7}$$

Considering the decay in the text, we have $A = 35$ and $Z = 18$: hence *the term multiplying* a_A *vanishes*. Inverting equation (1.7) we obtain

$$a_C = -\frac{A^{1/3} \Delta B}{2Z - 1}.$$

From (1.6) we have

$$\Delta B = [M_p - M_n - m]c^2 - Q_\beta = -1.293 - 0.511 - 4.95 \simeq -6.75 \text{ MeV},$$

where we have used the maximum positron energy for Q_β. Finally we get

$$a_C = -\frac{35^{1/3} \times -6.75}{35} \simeq 0.63 \text{ MeV}.$$

The value obtained in this way differs from the best-fit value (0.697 MeV) given with the *SEMF* by less than 10%.

Exercise 1.3.8

Denoting by Q_- the Q-value for

$$^{100}_{43}\text{Tc} \rightarrow {}^{100}_{44}\text{Ru} + e^- + \bar{\nu}_e$$

and with Q_+ the one for

$$^{100}_{43}\text{Tc} \rightarrow {}^{100}_{42}\text{Mo} + e^+ + \nu_e$$

we have (omitting c^2 multiplying the masses)

$$Q_- = 43M_p + 57M_n - B(100, 43) - 44M_p - 56M_n + B(100, 44) - m_e =$$

$$= M_n - M_p - m_e + B(100, 44) - B(100, 43) \simeq$$

$$\simeq B(100, 44) - B(100, 43) + 0.782 \text{ MeV}$$

and

$$Q_+ = M_p - M_n - m_e + B(100, 42) - B(100, 43) \simeq$$

$$\simeq B(100, 42) - B(100, 43) - 1.804 \text{ MeV}$$

From the *semi-empirical mass formula* we have for an odd-A odd-Z nucleus

$$B(A, Z) - B(A, Z \pm 1) = -a_C \frac{Z^2 - (Z \pm 1)^2}{A^{1/3}} -$$

$$-a_A \frac{(A - 2Z)^2 - [A - 2(Z \pm 1)]^2}{A} - 2\frac{a_P}{A^{1/2}}.$$

In the current case we have

$$B(100, 43) - B(100, 44) = -0.697 \times \frac{43^2 - 44^2}{100^{1/3}} -$$

$$-23.3 \times \frac{(100 - 86)^2 - (100 - 88)^2}{100} - 2 \times \frac{12}{\sqrt{100}} \simeq -1.45 \text{ MeV}$$

$$B(100, 43) - B(100, 42) = -0.697 \times \frac{43^2 - 42^2}{100^{1/3}} -$$

$$-23.3 \times \frac{(100 - 86)^2 - (100 - 84)^2}{100} - 2 \times \frac{12}{\sqrt{100}} \simeq -1.18 \text{ MeV}.$$

Hence we obtain

$$Q_- \simeq 2.23 \text{ MeV} \qquad Q_+ \simeq -0.62 \text{ MeV}.$$

The β^- decay to $^{100}_{44}$Ru is allowed. Instead the β^+ decay is forbidden, yet the electron capture, allowing the transition to $^{100}_{42}$Mo, is possible since we have $Q_{EC} = Q_+ + 2m_e \simeq 0.40 \text{ MeV} > 0$.

Exercise 1.3.9

(a) Each fission reaction releases 200 MeV $= 2 \ 10^8$ eV $\times 1.6 \ 10^{-19}$ J/eV $\simeq 3.2 \ 10^{-11}$ J. Hence the fission rate is

$$r = \frac{P}{E_{\text{fiss}}} = \frac{2 \ 10^9}{3.2 \ 10^{-11}} \simeq 6.25 \ 10^{19} \ \text{s}^{-1}.$$

(b) 1 g of ^{235}U releases an energy

$$E_{\text{fiss}} \times \frac{N_A}{A} = 3.2 \ 10^{-11} \times \frac{6 \ 10^{23}}{235} \simeq 0.8 \ 10^{11} \ \text{J/g}.$$

In a year the total energy is

$$2 \ 10^9 \ \frac{\text{J}}{\text{s}} \times 3.15 \ 10^7 \ \text{s} \simeq 6.3 \ 10^{16} \ \text{J}.$$

The ^{235}U mass consumed in a year is then

$$M(^{235}\text{U}) = \frac{6.3 \ 10^{16} \ \text{J}}{0.8 \ 10^{11} \ \text{J/g}} \simeq 7.88 \ 10^5 \ \text{g} = 788 \ \text{kg}.$$

Since ^{235}U is about 30%, the used fuel mass is about 2.6 ton.
(c) The maximum neutrino energy is equal to the Q-factor of the beta decay. Denoting it by Q_-, for the $^{145}_{57}$La β^- decay, we have (omitting c^2 in the mass terms):

$$Q_- = M_n - M_p - m_e - B(145, 57) + B(145, 58) \simeq 0.782 \ \text{MeV} - \Delta B_-$$

where ΔB_- is the difference in binding energy between parent and daughter nuclei. Using the *SEMF* we have for odd-A nuclei

$$\Delta B_- = B(A, Z) - B(A, Z + 1) \simeq$$

$$\simeq -a_C \frac{Z^2 - (Z + 1)^2}{A^{1/3}} - a_A \frac{(A - 2Z)^2 - [A - 2(Z + 1)]^2}{A},$$

which becomes in our case

$$\Delta B_- = -0.697 \times \frac{57^2 - 58^2}{145^{1/3}} - 23.3 \times \frac{(145 - 114)^2 - (145 - 116)^2}{145} \simeq -4.03 \ \text{MeV}$$

The maximum neutrino energy is then $0.782 + 4.03 \simeq 4.81$ MeV.
(d) The neutrino intensity is 20% of the fission rate $0.20 \times 6.25 \ 10^{19} \ \text{s}^{-1} = 1.25 \ 10^{19} \ \text{s}^{-1}$. At 500 m distance the neutrino flux is

$$\Phi = \frac{I_\nu}{4\pi R^2} \simeq \frac{1.25 \ 10^{19}}{12.56 \ 500^2} \simeq 4 \times 10^{12} \ \text{m}^{-2}\text{s}^{-1}$$

(e) For a detector having a length l (along the neutrino direction), a section S, composed of material of atomic mass A, the interaction rate is

$$r = \Phi \times \sigma \times \frac{N_A}{A} \times \rho l S = \Phi \times \sigma \times \frac{N_A}{A} \times M.$$

Such proportionality between rate and mass is holding each time the detector length is much smaller of the interaction length. Inserting our values we get

$$r = 4 \times 10^{12}\, \mathrm{m^{-2}s^{-1}} \times 6\,10^{-48}\mathrm{m^2} \times \frac{6\,10^{23}}{gA} \times 10^6\,\mathrm{g} \simeq \frac{1.44\,10^{-5}}{A}\,\mathrm{s^{-1}} \simeq \frac{450}{A}\mathrm{yr^{-1}}$$

Exercise 1.3.10

Let us call Q_+ the Q-factor for the β^+-decay of $^{27}_{14}\mathrm{Si}$. Omitting the factor c^2 multiplying the mass terms, we have:

$$(E_e)_{\max} = Q_+ = M_p - M_n - m_e - B(27, 14) + B(27, 13)$$

(a) The binding energy of $^{27}_{13}\mathrm{Al}$ is then

$$B(27, 14) = B(27, 13) + M_p - M_n - m_e - (E_e)_{\max} = 219.36\,\mathrm{MeV}$$

(b) The nuclei involved in the decay are odd-A, hence the pairing term of the *SEMF* disappears from the mass difference. The volume and surface terms do not contribute in any case. Considering the two surviving terms (a_C and a_A), the asymmetry term does not contribute since we have, for $A = 27$ and $Z = 14$, $(A - 2Z)^2 = (A - 2(Z - 1))^2$. Hence the mass difference depends only on the Coulomb term a_C.
(c) For a uniform charge distribution, we have

$$\Delta B = \frac{3}{5} \frac{e^2}{4\pi\epsilon_0 R}[Z^2(\mathrm{Si}) - Z^2(\mathrm{Al})] = \frac{3}{5} \frac{\alpha\hbar c}{R}[Z^2(\mathrm{Si}) - Z^2(\mathrm{Al})].$$

Hence we get for the $^{27}_{14}\mathrm{Si}$ radius

$$R = \frac{3}{5} \frac{\alpha\hbar c}{\Delta B}[Z^2(\mathrm{Si}) - Z^2(\mathrm{Al})] = 0.6 \times \frac{1.4\,\mathrm{MeV\,fm}}{5.59\,\mathrm{MeV}} \times (14^2 - 13^2) \simeq 4.1\,\mathrm{fm}$$

Exercise 1.3.11

The reaction $\nu_e + ^{71}_{31}\mathrm{Ga} \to ^{71}_{32}\mathrm{Ge} + e^-$ belongs to the more general class

$$\nu_e + (A, Z) \to (A, Z+1) + e^-,$$

for which the neutrino threshold energy is

$$E_{\mathrm{th}} = \frac{(m_e + M')^2 - M^2}{2M},$$

where M and M' are the masses of the nuclei (A, Z) and $(A, Z + 1)$ respectively. Denoting by ΔM the mass difference $M' - M$, we have

$$E_{th} = \frac{[m_e + (M + \Delta M)]^2 - M^2}{2M}$$

which, being $\Delta M, m_e \ll M$, becomes

$$E_{th} \simeq m_e + \Delta M. \tag{1.8}$$

We can write ΔM as

$$\Delta M = M' - M = M_p - M_n + \Delta B/c^2 \tag{1.9}$$

with

$$\Delta B = B(A, Z) - B(A, Z + 1).$$

Combining (1.8) and (1.9) we finally get

$$B(71, 32) = B(71, 31) - \Delta B = B(71, 31) - \Delta M - (M_n - M_p)c^2 =$$

$$= B(71, 31) - E_{th} - (M_n - M_p - m_e)c^2 \simeq$$

$$\simeq 618.95 - 0.233 - 0.782 \simeq 617.93 \text{ MeV}.$$

Using the *SEMF* we have instead

$$B(71, 32) \simeq 620.88 \text{ MeV}$$

which differs about 0.5 percent from the previous value.

Exercise 1.3.12

The minimum atomic mass for isobars can be obtained from the equation $\frac{\partial M(A, Z)}{\partial Z} = 0$. Using the *SEMF* we obtain

$$\frac{2a_C Z}{A^{1/3}} - \frac{4a_A(A - 2Z)}{A} + (M_n - M_p - m_e)c^2 = 0$$

The term which depends on the electromagnetic coupling constant is a_C. Solving in a_C we get

$$a_C = \frac{A^{1/3}}{2 Z} \left[\frac{4a_A(A - 2Z)}{A} - (M_n - M_p - m_e)c^2 \right]$$

If the stable nucleus is $^{133}_{54}\text{Xe}$, the a_C parameter should be

$$a_C = \frac{133^{1/3}}{2 \times 54} \left[\frac{4 \times 23.3\,(133 - 108)}{133} - 0.782 \right] \text{ MeV} \simeq 0.791 \text{ MeV}.$$

From classical electrostatics we know that the Coulomb term is proportional to the fine structure constant $a_C \propto \alpha$. Hence the change in the coupling constant is

$$\frac{\Delta\alpha}{\alpha} = \frac{\Delta a_C}{a_C} \simeq \frac{0.791 - 0.697}{0.697} \simeq 13\%$$

Exercise 1.3.13

(a) For photons colliding against a fixed iron target, the threshold energy is

$$E_\gamma^{th} = \frac{(M' + m)^2 - M^2}{2M} = \frac{\mu^2}{2M} \tag{1.10}$$

where M is the mass of the initial nucleus, (A, Z), M' that of the final nucleus, $(A - 1, Z)$, and m the neutron mass. Denoting by ΔM the nuclear mass difference $M - M'$ and with ΔB the corresponding binding energy difference, we have

$$\mu^2 = [(M - \Delta M) + m]^2 - M^2 = m(m + 2M) - \Delta M(2M + 2m - \Delta M).$$

Since $\Delta M, m \ll M$, we get

$$\mu^2 \simeq 2M(m - \Delta M) = 2M\Delta B, \tag{1.11}$$

having used the relation

$$\Delta M = M - M' = m - \Delta B$$

For $^{56}_{26}$Fe photo-disintegrating into $^{55}_{26}$Fe, ΔB can be obtained from the *semi-empirical mass formula*

$$\Delta B = B(56, 26) - B(55, 26) \simeq 490.95 - 478.90 \simeq 12 \text{ MeV}. \tag{1.12}$$

Hence the photon threshold energy is

$$E_\gamma^{th} \simeq \frac{2M\Delta B}{2M} = \Delta B \simeq 12 \text{ MeV}.$$

(b) In the case of Cosmic Rays, the collision does not occur in a fixed target frame and the expression (1.10) cannot be used. Instead we make use of the invariance of the total 4-momentum squared so that we can write, at the threshold

$$(M' + m)^2 = (E_N^{th} + E_\gamma)^2 - (\mathbf{p}_N^{th} + \mathbf{p}_\gamma)^2 = E_N^{th^2} + E_\gamma^2 + 2E_N^{th}E_\gamma - \mathbf{p}_N^{th^2} - \mathbf{p}_\gamma^2 - 2\mathbf{p}_N^{th} \cdot \mathbf{p}_\gamma,$$

where $(E_N^{th}, \boldsymbol{p}_N^{th})$ is the 4-momentum of the initial nucleus (at the threshold energy) and $(E_\gamma, \boldsymbol{p}_\gamma)$ the photon 4-momentum. Since nuclei are ultra-relativistic we have

$$(M' + m)^2 \simeq M^2 + 2E_N^{th}E_\gamma(1 - \cos\theta),$$

where θ is the angle between the nucleus and photon directions. Using μ^2 we obtain

$$2E_N^{th}E_\gamma(1 - \cos\theta) \simeq \mu^2.$$

The nucleus threshold energy is

$$E_N^{th} = \frac{\mu^2}{2E_\gamma(1 - \cos\theta)}.$$

Using (1.11) and $\theta = \pi$ (head-on collisions), we finally get

$$E_N^{th} = \frac{M\Delta B}{2E_\gamma} = \frac{52 \text{ GeV } 12 \text{ MeV}}{2 \; 10^{-3} \text{ eV}} \simeq 3 \times 10^{20} \text{ eV}$$

where we have used $M = M(56, 26) \simeq 52$ GeV, as obtained from the *SEMF*.

Exercise 1.3.14

(a) We can write the two separation energies as

$$S_p = B(A, Z) - B(A - 1, Z - 1)$$

$$S_n = B(A, Z) - B(A - 1, Z)$$

hence we obtain for the difference

$$S_p - S_n = B(A - 1, Z) - B(A - 1, Z - 1).$$

Using the *SEMF* we get

$$S_p - S_n = -a_C \frac{Z^2 - (Z - 1)^2}{(A - 1)^{1/3}} - a_A \frac{(A - 1 - 2Z)^2 - [A - 1 - 2(Z - 1)]^2}{A - 1} + D_P =$$

$$- a_C \frac{2Z - 1}{(A - 1)^{1/3}} + 4a_A \frac{A - 2Z}{A - 1} + D_P, \tag{1.13}$$

where D_P originates from the difference of the *pairing* terms, $\delta_P(A)$.

The possible values of D_P are reported in the following table, where e and o stand for even and odd nucleon parity in the corresponding nucleus.

(b) If in Eq. (1.13) we insert $Z = A/2$, valid for light nuclei, we obtain

parent	S_p : $(Z, N-1)$		S_n : $(Z-1, N)$		D_P
Z, N	parity	δ_P	parity	δ_P	
e e	e o	0	o e	0	0
o o	o e	0	e o	0	0
e o	e e	$+a_P/\sqrt{A-1}$	o o	$-a_P/\sqrt{A-1}$	$+2a_P/\sqrt{A-1}$
o e	o o	$-a_P/\sqrt{A-1}$	e e	$+a_P/\sqrt{A-1}$	$-2a_P/\sqrt{A-1}$

$$S_p - S_n = -a_C \frac{A-1}{(A-1)^{1/3}} = -a_C(A-1)^{2/3}.$$

Note that in this case we have $D_P = 0$, because $A = 2Z$ is necessarily even, corresponding to the first two rows of the table. The difference $S_p - S_n$ is always negative and decreasing with A. Hence larger energy is needed to extract neutrons than protons.

(c) If instead we use $Z = A/2.5$, approximately valid for heavy nuclei, we obtain

$$S_p - S_n = -a_C \frac{2/2.5\,A - 1}{(A-1)^{1/3}} + a_A \frac{2}{2.5} \frac{A}{A-1} + D_P =$$

$$= \frac{2}{2.5}\left[-a_C \frac{A - 1.25}{(A-1)^{1/3}} + a_A \frac{A}{A-1}\right] + D_P.$$

Also in this case the difference $S_p - S_n$ decreases with A. The general treatment is complicated because of the presence of D_P, which can have either sign.

For even-A nuclei ($D_P = 0$) the curve starts from positive values because of the a_A term. In the following figure the $S_p - S_n$ behaviour is shown for even-A nuclei.

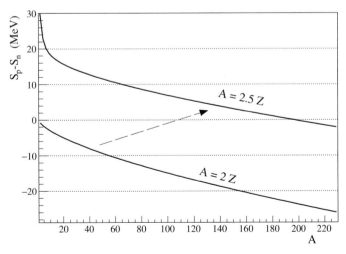

(d) Using (1.13) we get

- $^{20}_{10}$Ne: $S_p - S_n \simeq -4.96$ MeV;
- $^{38}_{18}$Ar: $S_p - S_n \simeq -2.28$ MeV;
- $^{106}_{46}$Pd: $S_p - S_n \simeq -1.02$ MeV;
- $^{137}_{56}$Ba: $S_p - S_n \simeq +4.15$ MeV.
- $^{200}_{80}$Hg: $S_p - S_n \simeq -0.248$ MeV.

Apart the fourth nucleus, $^{137}_{56}$Ba, all nuclei are even A and the values of the separation differences can be interpreted looking at the figure. The nuclei reported in the list increase with A and show a transition from case (a) to case (b). The two assumptions are strictly valid only for the first and last nuclei. The arrow in the figure sketches the transition region.

1.4 Nuclear Decays

Exercise 1.4.1

Denoting by T_α the kinetic energy of the emitted α-particle, we have approximately

$$Q_\alpha \simeq \frac{A}{A-4} T_\alpha.$$

Hence the Q_α values are

$$Q_1 \simeq \frac{240}{236} \times 5.17 \simeq 5.26 \text{ MeV}$$

$$Q_2 \simeq \frac{240}{236} \times 5.12 \simeq 5.21 \text{ MeV}$$

The γ energy corresponds to the difference between Q_1 (^{240}Pu $\rightarrow$ ^{236}U) and Q_2 (^{240}Pu $\rightarrow$ ^{236}U*). Hence we get

$$E_\gamma = Q_1 - Q_2 = 0.05 \text{ MeV}$$

Exercise 1.4.2

Calling τ_1 and N_1 the mean lifetime and number of ^{244}Pu nuclei at time t, τ_2 and N_2 the same quantities for ^{240}U and τ_3 and N_3 the same quantities for ^{240}Np, we have $\tau_1 \gg \tau_2, \tau_3$. Furthermore if t corresponds to the time of the measurement ($= 30$ d), we have also $t \ll \tau_1$. Under these conditions the *secular equilibrium* equation[2] holds

$$\frac{N_1}{\tau_1} \simeq \frac{N_2}{\tau_2} \simeq \frac{N_3}{\tau_3}.$$

---

[2] See e.g. problem 1.4.6. The secular equilibrium is obtained for $\omega_1 \ll \omega_2, \omega_3$ and $\omega_1 t \ll 1$.

The source has a mass of 1 mol and then $N_1 = N_A$, with N_A the Avogadro number

$$N(^{240}U) = N_2 = \frac{\tau_2}{\tau_1} N_1 = \frac{T_2^{(1/2)}}{T_1^{(1/2)}} N_A \simeq \frac{14\,h \times 6.02 \cdot 10^{23}}{8.1 \cdot 10^7 \times 365 \times 24\,h} \simeq 1.2 \cdot 10^{13}$$

$$N(^{240}Np) = N_3 = \frac{\tau_3}{\tau_1} N_1 = \frac{T_3^{(1/2)}}{T_1^{(1/2)}} N_A \simeq$$

$$\simeq \frac{67\,min \times 6.02 \cdot 10^{23}}{8.1 \cdot 10^7 \times 365 \times 24 \times 60\,min} \simeq 9.4 \cdot 10^{11}$$

The decays involved in the chain are

$$^{244}_{94}Pu \rightarrow {}^{240}_{92}U + \alpha$$

$$^{240}_{92}U \rightarrow {}^{240}_{93}Np + e^- + \bar{\nu}_e$$

$$^{240}_{93}Np \rightarrow {}^{240}_{94}Pu + e^- + \bar{\nu}_e.$$

The activity measured in the α-decay is

$$\mathcal{A} = \left| \frac{dN_1}{dt} \right| \simeq \frac{N_1}{\tau_1} = \frac{N_1 \ln 2}{T_1^{(1/2)}} \simeq \frac{6.02 \cdot 10^{23} \times 0.69}{8.1 \cdot 10^7 \times 365 \times 24 \times 3600\,s} \simeq 1.6 \cdot 10^8\,s^{-1}.$$

Exercise 1.4.3

The decay constant of ^{226}Ra is

$$\omega = \frac{\ln 2}{T_{1/2}} \simeq \frac{0.693}{1.6 \times 10^3 \times 3.15 \times 10^7\,s} \simeq 1.4\,10^{-11} s^{-1}.$$

The activity of a source is given by

$$\mathcal{A}(t) = \left| \frac{dN}{dt} \right| = \omega N(t) = \omega N_0 e^{-\omega t} \qquad [s^{-1}]$$

The number of ^{226}Ra nuclei at time 0 is

$$N_0 = \frac{N_A}{A} = \frac{6.02 \times 10^{23}}{226} \simeq 2.66 \times 10^{21}$$

Hence we have

$$\mathcal{A}(0) = \omega N_0 \simeq 1.4\,10^{-11} \times 2.66\,10^{21} \simeq 3.7\,10^{10}\,s^{-1}.$$

Note that this is the definition of 1 Curie (1 Ci).

Exercise 1.4.4

The $^{14}_{6}C$ beta decay is $^{14}_{6}C \rightarrow {}^{14}_{7}N + e^- + \bar{\nu}_e$. The specimen activity is given by

$$\mathcal{A} = \left| \frac{dN}{dt} \right| = \frac{N(^{14}C)}{\tau(^{14}C)}.$$

The number of $^{14}_{6}C$ nuclei present in the specimen when it was still a living organism is

$$N_0(^{14}C) = f \times N_0(C) = f \times m \times \frac{N_A}{\langle A(C) \rangle} \simeq 1.3 \cdot 10^{-12} \times 5 \times \frac{6.02 \cdot 10^{23}}{12.001} \simeq 3.3 \cdot 10^{11},$$

where f is the fraction of $^{14}_{6}C$ nuclei in a living organism, m its mass, N_A the Avogadro number and $\langle A(C) \rangle$ the atomic mass of natural carbon. The $^{14}_{6}C$ mean lifetime is $\tau(^{14}C) = T_{1/2}(^{14}C)/\ln 2 \simeq 8200$ years. Hence we have for the specimen activity when the organism died

$$\mathcal{A}_0 = \frac{N_0(^{14}C)}{\tau(^{14}C)} \simeq \frac{3.3 \cdot 10^{11}}{8200 \times 3.15 \cdot 10^7 \text{ s}} \simeq 1.28 \text{ s}^{-1},$$

The present activity is related to $\mathcal{A}_0$ through the equation

$$\mathcal{A}(t) = \mathcal{A}_0 \cdot e^{-t/\tau(^{14}C)} = \frac{3600}{2 \times 3600 \text{ s}} \simeq 0.5 \frac{\text{decays}}{\text{s}},$$

hence we get the age of the fossil

$$T = -\tau(^{14}C) \times \ln \frac{\mathcal{A}(t)}{\mathcal{A}_0} \simeq -8200 \text{ yr} \times \ln \frac{0.5}{1.28} \simeq 7700 \text{ yr.}$$

Exercise 1.4.5

The nucleus ^{226}Ra has a decay constant given by

$$\omega = \frac{\ln 2}{T_{1/2}} = \frac{0.693}{1.6 \times 10^3 \times 3.15 \times 10^7 \text{s}} \simeq 1.4 \ 10^{-11} \text{ s}^{-1}.$$

For 1 g of ^{226}Ra we have then an activity

$$\mathcal{A} = \left| \frac{dN}{dt} \right| = \omega N_0 = \omega \frac{N_A}{A} \times 1 \simeq 1.4 \ 10^{-11} \times \frac{6.02 \times 10^{23}}{226} \simeq 3.7 \ 10^{10} \text{ s}^{-1}.$$

This is the current definition of 1 Curie (1 Ci).

The ^{60}Co source we are considering has an activity of 10 Ci, that is $3.7 \ 10^{11} \ \text{s}^{-1}$. If m is its mass we get

$$m = \mathcal{A} \frac{A}{N_A} \frac{T_{1/2}}{\ln 2} \simeq 3.7 \ 10^{11} \times \frac{60}{6.02 \times 10^{23}} \ \frac{5.26 \times 3.15 \times 10^7}{0.693} \simeq 8.8 \ \text{mg}.$$

A simpler approach to get the same result is obtained using the following relation, which holds for *sources with equal activities*

$$\frac{m_1}{m_2} = \frac{A_1}{A_2} \times \frac{T_{1/2}^{(1)}}{T_{1/2}^{(2)}}.$$

This equation can be used in our case knowing that our source has the same activity of $10 \, \text{g}$ of ^{226}Ra. Hence we get

$$m_{Co} = m_{Ra} \times \frac{A_{Co}}{A_{Ra}} \times \frac{T_{1/2}^{(Co)}}{T_{1/2}^{(Cu)}} \simeq 10 \times \frac{60}{226} \times \frac{5.26}{1600} \simeq 8.7 \ \text{mg}.$$

Exercise 1.4.6

The numbers of nuclei of the three types are ruled by the following nested equations

$$\frac{dN_1}{dt} = -\omega_1 N_1$$

$$\frac{dN_2}{dt} = \omega_1 N_1 - \omega_2 N_2$$

$$\frac{dN_3}{dt} = \omega_2 N_2 - \omega_3 N_3.$$

In our case, the initial conditions are $N_1(0) = N_0$, $N_k(0) = 0$ and $dN_k/dt(0) = 0$ for $k = 2, 3$. The particular solution for these consitions is

$$N_1(t) = N_0 e^{-\omega_1 t}$$

$$N_2(t) = N_0 \frac{\omega_1}{\omega_2 - \omega_1} (e^{-\omega_1 t} - e^{-\omega_2 t})$$

$$N_3(t) = N_0 \omega_1 \omega_2 \left[\frac{e^{-\omega_1 t}}{(\omega_2 - \omega_1)(\omega_3 - \omega_1)} + \frac{e^{-\omega_2 t}}{(\omega_3 - \omega_2)(\omega_1 - \omega_2)} + \frac{e^{-\omega_3 t}}{(\omega_1 - \omega_3)(\omega_2 - \omega_3)} \right].$$

uhe nucleus 3 is stable and then $\omega_3 = 0$ and $N_3(t)$ can be written as

$$N_3(t) = N_0 \left[1 + \frac{e^{-\omega_1 t}}{\omega_1/\omega_2 - 1} + \frac{e^{-\omega_2 t}}{\omega_2/\omega_1 - 1} \right].$$

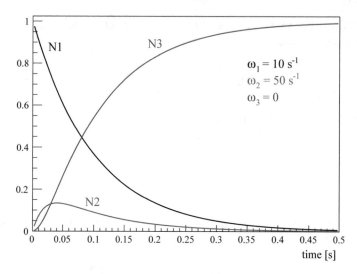

Fig. 1.3 Relative abundances for a decay chain with three nuclei having the decay constants given in the figure

In Fig. 1.3 the three nuclear populations are shown as a function of time for the decay constants given in the text. For $t = 1/4$ s we obtain

$$\frac{N_3}{N_1} = \frac{\left[1 + \frac{e^{-\omega_1 t}}{\omega_1/\omega_2 - 1} + \frac{e^{-\omega_2 t}}{\omega_2/\omega_1 - 1}\right]}{e^{-\omega_1 t}} \simeq 10.9.$$

Exercise 1.4.7

The fraction of ^{238}U isotopes decayed in $2.5 \ 10^9$ years is

$$f = 1 - \exp\left(-\frac{t}{\tau}\right) = 1 - \exp\left(-\frac{t \ln 2}{T_{1/2}}\right) \simeq$$

$$\simeq 1 - \exp\left(-\frac{2.5 \ 10^9 \times 0.693}{4.5 \ 10^9}\right) \simeq 32\%$$

The specific activity is the activity per unit mass. Hence we have

$$a = \frac{\mathcal{A}}{M} = \frac{N_A}{A} \frac{\ln 2}{T_{1/2}} \simeq$$

$$\simeq \frac{6.02 \ 10^{23}}{238} \frac{0.693}{4.5 \ 10^9 \times 3.15 \ 10^7} \simeq 1.23 \ 10^4 \ \text{s}^{-1} \cdot \text{g}^{-1} \simeq 0.33 \ \frac{\mu\text{Ci}}{\text{g}}$$

Exercise 1.4.8

1. $^{44}_{22}\text{Ti} \rightarrow \ ^{40}_{20}\text{Ca} + \alpha.$

This decay is not allowed. Only nuclei having $A \geq 200$ can fulfill the kinematical conditions for the α-decay.

2. $^{241}_{95}\text{Am} \rightarrow ^{237}_{93}\text{Np} + \alpha$.

This decay is allowed if $Q_\alpha > 0$. Using the $SEMF$ we obtain

$$Q_\alpha = M(241, 95) - M(237, 93) - M_\alpha = B(237, 93) - B(241, 95) + B_\alpha$$

$$= 1798 - 1820 + 28.3 \simeq 5.91 \text{ MeV}.$$

Hence the decay is allowed.

3. $^{141}_{55}\text{Cs} \rightarrow ^{141}_{56}\text{Ba} + e^+ + \nu_e$.

This decay is forbidden because charge is not conserved.

4. $^{69}_{28}\text{Ni} \rightarrow ^{69}_{29}\text{Cu} + e^- + \bar{\nu}_e$.

This decay is allowed, provided that we have $Q_{\beta-} > 0$.

$$Q_{\beta-} = M(69, 28) - M(69, 29) - m_e$$

$$= 28M_p + 41M_n - B(69, 28) - 29M_p - 40M_n + B(69, 29) - m_e$$

$$= M_n - M_p - m_e + B(69, 29) - B(69, 28) \simeq 0.782 + 600.0 - 593.5 \simeq 7.3 \text{ MeV}$$

Hence the decay is allowed.

Exercise 1.4.9

As in problem 1.4.6, the time evolution of the nuclei involved in the chain is

$$N_1(t) = N_0 e^{-\omega_1 t}$$

$$N_2(t) = N_0 \frac{\omega_1}{\omega_2 - \omega_1} (e^{-\omega_1 t} - e^{-\omega_2 t})$$

$$N_3(t) = N_0 \omega_1 \omega_2 \left[\frac{e^{-\omega_1 t}}{(\omega_2 - \omega_1)(\omega_3 - \omega_1)} + \frac{e^{-\omega_2 t}}{(\omega_3 - \omega_2)(\omega_1 - \omega_2)} + \frac{e^{-\omega_3 t}}{(\omega_1 - \omega_3)(\omega_2 - \omega_3)} \right].$$

In our case the third equation is not used. We have $\omega_1 = \ln 2/2.25 = 0.31$ min^{-1}, $\omega_2 = \ln 2/22.9 = 0.03$ min^{-1}. The maximum N_2 is found solving the equation

$$\frac{d N_2}{dt} = 0$$

$$-\omega_1 e^{-\omega_1 t} + \omega_2 e^{-\omega_2 t} = 0$$

whose solution is

$$t = \frac{\ln(\omega_1/\omega_2)}{\omega_1 - \omega_2} \simeq 8.5 \text{ min}$$

The time dependence of the nuclear fractions is shown below.

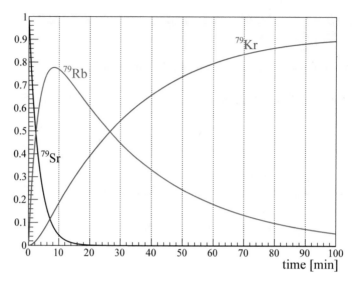

Exercise 1.4.10

Denoting by N_U and N_{Rn} the numbers of ^{238}U and ^{222}Rn nuclei, ω_U and ω_{Rn} their decay constants, the condition of *secular equilibrium* can be written as

$$N_U \, \omega_U \; = \; N_{Rn} \, \omega_{Rn}.$$

ω_{Rn} is related to the specific activity measurement as

$$a = \frac{\mathscr{A}}{V} = \frac{N_{Rn} \, \omega_{Rn}}{V} = \frac{N_U \, \omega_U}{V}$$

where V is the volume of the basement (60 m^3). N_U can be expressed as the ^{238}U concentration ρ_U times the volume from which the Radon gas diffuses

$$a = \frac{\rho_U \, S \, d \, \omega_U}{V},$$

where S is the surface of the walls (94 m^2) and d is the thickness (0.02 m) of the layer from which the gas diffuses. Hence we have

$$\rho_U = \frac{a \, V}{S \, d \, \omega_U} = \frac{a \, V \, T_U^{1/2}}{S \, d \, \ln 2} \simeq \frac{100 \times 60 \times 4.5 \; 10^9 \times 3.15 \; 10^7}{94 \times 0.02 \times 0.693 \text{ m}^3} \simeq 6.5 \times 10^{20} \text{ m}^{-3}$$

Exercise 1.4.11

The Q_α-value of the decay can be obtained from the alpha decay energy

$$Q_\alpha = \frac{A}{A-4} \, T_\alpha = \frac{239}{235} \, 5.144 \simeq 5.232 \text{ MeV}.$$

The measured power is equal to the intensity of the alpha decays multiplied by the released energy (Q_α). Hence we get

$$I = \frac{W}{Q_\alpha} \simeq \frac{0.231 \text{ Joule/s}}{5.232 \ 10^6 \text{ eV}} \simeq \frac{0.231 \text{ Joule/s}}{5.232 \ 10^6 \ 1.6 \ 10^{-19} \text{Joule}} \simeq 2.76 \ 10^{11} \text{ s}^{-1}.$$

The half-life is then

$$T_{1/2} = \frac{N(^{239}\text{Pu}) \ \ln 2}{I} = \frac{N_A \ m(^{239}\text{Pu}) \ \ln 2}{A \ I} \simeq$$

$$\simeq \frac{6.02 \ 10^{23} \times 120 \times 0.693}{239 \times 2.76 \ 10^{11}} \simeq 7.57 \ 10^{11} \text{ s}^{-1} \simeq 24000 \text{ yr}.$$

1.5 Nuclear Models

Exercise 1.5.1

The Saxon-Woods potential has the following expression

$$V(r) = \frac{-V_0}{1 + \exp \frac{r-R}{d}},$$

where $-V_0$, R and d are the three potential parameters, representing respectively the minimum depth, the nuclear radius and the thickness of region where nuclear matter vanishes.

Taken any spherical potential well, a larger radius generates eigenfunctions which are contained in larger volumes. As a consequence the energy levels (i.e. the eigenvalues) decrease. Hence for a nucleus with larger radius we expect lower energy levels.

A more quantitative result cannot be obtained, as the radial Schödinger equation for a Saxon-Woods potential is not analytically integrable. However using the Fermi gas model we can estimate the relative effect. The Fermi energy is:

$$E_F = \frac{1}{2m} \left(\frac{\hbar}{r_0} \right)^2 \left(\frac{9\pi}{8} \right)^{2/3} \simeq 33 \text{MeV}, \tag{1.14}$$

where m is the nucleon mass and $r_0 \simeq 1.2$ fm is the coefficient of the nuclear radius A-dependence ($R \simeq r_0 A^{1/3}$). This energy represents the maximum kinetic energy of nucleons in the nucleus. The energy of the ground state is obtained as $E_{GS} = -V_0 + E_F$, with $V_0 \approx 41$ MeV to agree with a binding energy per nucleon of about 8 MeV. From (1.14) increasing by 50% the nuclear radius one gets

$$E_{GS} = -V_0 + E_F(1.5 \ r_0) \simeq -41 + 15 \simeq -26 \text{ MeV},$$

which has to be compared with -8 MeV for the standard radius.

The binding energy is related to the ground state energy (about equal to its absolute value), hence it increases with the nuclear radius.

Exercise 1.5.2

The carbon isotopes have 6 protons. These are all contained in fully closed shells according to the configuration $(1s_{1/2})^2(1p_{3/2})^4$. Hence they do not contribute to the spin-parities of the nuclei. These are instead determined by the last neutron shells. The configurations of the carbon isotopes are

$^{11}C : (1s_{1/2})^2(1p_{3/2})^3$

$^{12}C : (1s_{1/2})^2(1p_{3/2})^4$

$^{13}C : (1s_{1/2})^2(1p_{3/2})^4(1p_{1/2})^1$

$^{14}C : (1s_{1/2})^2(1p_{3/2})^4(1p_{1/2})^2$

In all cases the last shell has $l = 1$. We then have for odd-N isotopes

$^{11}C : J = 3/2, P = (-1)^1 = - \qquad \Rightarrow J^P = 3/2^-$

$^{13}C : J = 1/2, P = (-1)^1 = - \qquad \Rightarrow J^P = 1/2^-$.

Instead for even-N isotopes, all neutrons are paired and then

$^{12}C, {}^{14}C : J = 0, P = + \qquad \Rightarrow J^P = 0^+$

Exercise 1.5.3

The first two nuclei are odd-A. Hence their spin and parity is determined by the last unpaired nucleon. The shell configurations are

$^{33}_{16}S \ n : (1s_{1/2})^2(1p_{3/2})^4(1p_{1/2})^2(1d_{5/2})^6(2s_{1/2})^2(1d_{3/2})^1$

$^{39}_{19}K \ p : (1s_{1/2})^2(1p_{3/2})^4(1p_{1/2})^2(1d_{5/2})^6(2s_{1/2})^2(1d_{3/2})^3$

Both nucleons have $l = 2$. Their spin and parity are

$^{33}_{16}S \ n : J = 3/2, P = (-1)^2 = + \qquad \Rightarrow J^P = 3/2^+$

$^{39}_{19}K \ p : J = 3/2, P = (-1)^2 = + \qquad \Rightarrow J^P = 3/2^+$

$^{64}_{28}Ni$ is an even-even nucleus, all nucleons are paired and then $J^P = 0^+$.

Exercise 1.5.4

Using the Fermi gas distribution we have

$$\langle E_k \rangle = \frac{\int_0^{p_F} \frac{p^2}{2M} d^3 p}{\int_0^{p_F} d^3 p} = \frac{4\pi}{2M} \frac{\int_0^{p_F} p^4 dp}{\int_0^{p_F} 4\pi p^2 dp} = \frac{3}{5} \frac{p_F^2}{2M}$$

where $p_F = \frac{\hbar}{2r_0}(9\pi)^{1/3}$ is the Fermi momentum ($r_0 = 1.2$ fm) and M is the nucleon mass (can be assumed equal for the purpose). Multiplying and dividing by c^2, we obtain

$$\langle E_k \rangle = \frac{3(\hbar c)^2(9\pi)^{2/3}}{40 \, r_0^2 M c^2} \simeq \frac{3 \times 197^2 \times 28.27^{2/3}}{40 \times 1.2^2 \times 940} \simeq 20 \text{ MeV}.$$

This expression does not depend on the content of protons (Z) and neutrons (N). Therefore the mean kinetic energy is the same for all nuclei.

Exercise 1.5.5

The nuclear shells involved and the spin-parity of the ground states are

- $^{15}_{7}$N odd-A nucleus, $p : (1s_{1/2})^2(1p_{3/2})^4(1p_{1/2})^1 \Rightarrow J^P = \frac{1}{2}^-$
- $^{27}_{12}$Mg odd-A nucleus, $n : (1s_{1/2})^2(1p_{3/2})^4(1p_{1/2})^2(1d_{5/2})^6(2s_{1/2})^1 \Rightarrow J^P = \frac{1}{2}^+$
- $^{60}_{28}$Ni even-even nucleus $\Rightarrow J^P = 0^+$
- $^{87}_{38}$Sr odd-A nucleus, $n : (1s_{1/2})^2(1p_{3/2})^4(1p_{1/2})^2(1d_{5/2})^6(2s_{1/2})^2(1d_{3/2})^4(1f_{7/2})^8$
  $(2p_{3/2})^4(1f_{5/2})^6(2p_{1/2})^2(1g_{9/2})^9 \Rightarrow J^P = \frac{9}{2}^+$

Exercise 1.5.6

All these isotopes can be unstable because of beta decay. Gamma decay is not possible because they are in the ground states. Alpha decay is kinematically forbidden for $A \lesssim 200$. To establish if they are stable it is then necessary to evaluate their Q_β values.

$^{15}_{8}$O nuclide. We have to calculate Q_-, Q_+ and Q_{EC} respectively for β^-, β^+ decays and electron capture (EC). We have

$$Q_- = -\Delta B_- + 0.782 \text{ MeV}$$

$$Q_+ = -\Delta B_+ - 1.804 \text{ MeV}$$

$$Q_{EC} = Q_+ + 1.022 \text{ MeV}$$

where $\Delta B_\mp = B(A, Z) - B(A, Z \pm 1)$, the difference between the binding energies of the parent and daughter nuclei, can be derived from the *SEMF*.

We obtain $Q_- < 0$, instead $Q_+ = 2.44$ MeV and $Q_{EC} = 3.46$ MeV. $^{15}_{8}$O is then unstable and can decay by both β^+-decay and EC.

This isotope is odd-A, so the spin and parity are determined by the unpaired neutron of the last shell. The neutron shell configuration and spin-parity are $n : (1s_{1/2})^2(1p_{3/2})^4(1p_{1/2})^1 \Rightarrow J^P = 1/2^-$.

In the shell model the magnetic moment is $\mu = g_J J$. We have an unpaired neutron ($g_l = 0$, $g_s = -3.83$ n.m.) and g_J is given by

$$g_J = g_s \frac{j(j+1) - l(l+1) + s(s+1)}{2j(j+1)} \tag{1.15}$$

which for $^{15}_{8}$O turns out to be $g_J = 1.28$ corresponding to a magnetic moment $\mu \simeq 0.64$ n.m.

$^{16}_{8}$O nuclide. This nucleus is even-even. Furthermore it is the well-known stable and most abundant oxygen isotope. Hence we have $J^P = 0^+$ and $\mu = 0$.

$^{17}_{8}$O nuclide. Evaluating Q_-, Q_+ e Q_{EC} as for the first nuclide we obtain negative values for all of them. Hence $^{17}_{8}$O is stable.

The same nuclide is odd-A. Again the spin and parity are determined by the unpaired neutron of the last shell. The shell configuration and spin-parity are $n : (1s_{1/2})^2(1p_{3/2})^4(1p_{1/2})^2(1d_{5/2})^1 \Rightarrow J^P = 5/2^+$.

Using Eq. (1.15) we now get $g_J = -0.77$ corresponding to a magnetic moment $\mu \simeq -1.92$ n.m.

Exercise 1.5.7

We have odd-A nuclei and so the spin and parity of the ground states are that of the unpaired nucleon. The maximum occupation is 15, corresponding to the neutron number for $^{29}_{14}$Si. The shell sequence up to 20 in the *standard* shell model, that is with *inverse* spin-orbit coupling, is

$$\text{(a)} \quad (1s_{1/2})^2 \, (1p_{3/2})^4 \, (1p_{1/2})^2 \, (1d_{5/2})^6 \, (2s_{1/2})^2 \, (1d_{3/2})^4.$$

If instead the spin-orbit coupling were *direct* (case b) the shell sequence proceeds with increasing J values as

$$\text{(b)} \quad (1s_{1/2})^2 \, (1p_{1/2})^2 \, (1p_{3/2})^4 \, (1d_{3/2})^4 \, (2s_{1/2})^2 \, (1d_{5/2})^6.$$

For $^7_3$Li, whose unpaired nucleon is a proton, we have

$$\text{(a)} \quad p : (1s_{1/2})^2(1p_{3/2})^1 \quad J^P = 3/2^-$$

$$\text{(b)} \quad p : (1s_{1/2})^2(1p_{1/2})^1 \quad J^P = 1/2^-.$$

For $^{29}_{14}$Si a neutron is unpaired and we have

$$\text{(a)} \quad n : (1s_{1/2})^2(1p_{3/2})^4(1p_{1/2})^2(1d_{5/2})^6(2s_{1/2})^1 \qquad\qquad J^P = 1/2^+$$

$$\text{(b)} \quad n : (1s_{1/2})^2(1p_{1/2})^2(1p_{3/2})^4(1d_{3/2})^4(2s_{1/2})^2(1d_{5/2})^1 \quad J^P = 5/2^+$$

Exercise 1.5.8

^{52}Cr is even-even and then spin-parity is $J^P = 0^+$. The other Cr isotopes are odd-A and J^P is that of the unpaired nucleon. The proton number is even, hence only the neutron shell configuration is relevant. We have
^{51}Cr   27 n : $(1s_{1/2})^2(1p_{3/2})^4(1p_{1/2})^2(1d_{5/2})^6(2s_{1/2})^2(1d_{3/2})^4(1f_{7/2})^7$
^{55}Cr   31 n : $(1s_{1/2})^2(1p_{3/2})^4(1p_{1/2})^2(1d_{5/2})^6(2s_{1/2})^2(1d_{3/2})^4(1f_{7/2})^8(2p_{3/2})^3.$
Therefore spin and parity of the ground states are
^{51}Cr $J = 7/2, P = (-1)^3 \Rightarrow J^P = 7/2^-$
^{55}Cr $J = 3/2, P = (-1)^1 \Rightarrow J^P = 3/2^-$
    These two isotopes ate unstable because of β decay. To find the possible decay modes we calculate Q_-, Q_+ e Q_{EC} respectively for β^-, β^+ and electronic capture (EC). These are

$$Q_- = -\Delta B_- + 0.782 \text{ MeV}$$

$$Q_+ = -\Delta B_+ - 1.804 \text{ MeV}$$

$$Q_{EC} = Q_+ + 1.022 \text{ MeV},$$

where $\Delta B_\mp = B(A, Z) - B(A, Z \pm 1)$ is the binding energy difference corresponding to each decay. To get $\Delta B_\mp$ we use the *SEMF*.

For the $^{51}_{24}$Cr isotope, Q_+ and Q_- are both negative, yet we find $Q_{EC} \simeq 1.52 - 1.804 + 1.022 \simeq 0.74$ MeV. This means that $^{51}_{24}$Cr transmutes to $^{51}_{23}$V by electron capture.

For $^{55}_{24}$Cr, Q_+ and Q_{EC} are both negative, but $Q_- \simeq 1.19 + 0.782 \simeq 1.97$ MeV. Hence this nucleus decays to $^{55}_{25}$Mn by β^--decay.

Exercise 1.5.9

(a) ^{57}Cu e ^{57}Ni are *mirror* nuclei with a single nucleon (valence nucleon) out of complete shells. The valence nucleon is a proton for ^{57}Cu and a neutron for ^{57}Ni. The shell sequence is

$$1s_{1/2} \; 1p_{3/2} \; 1p_{1/2} \; 1d_{5/2} \; 2s_{1/2} \; 1d_{3/2} \; 1f_{7/2} \; 2p_{3/2} \; 1f_{5/2} \; ...$$

The shell configuration up to the valence nucleon (occupancy no. 29) is

$$(1s_{1/2})^2(1p_{3/2})^4(1p_{1/2})^2(1d_{5/2})^6(2s_{1/2})^2(1d_{3/2})^4(1f_{7/2})^8(2p_{3/2})^1$$

and the first excited level corresponds to the following shell, $1f_{5/2}$.

Hence we have for spin and parity

$$\text{GS}: l = 1, \; j = \frac{3}{2} \Rightarrow J^P = \frac{3}{2}^- \; ; \quad \text{1st Exc}: l = 3, \; j = \frac{5}{2} \Rightarrow J^P = \frac{5}{2}^- .$$

(b) The magnetic moment is $\mu = g_j \, j$, where

$$g_j = g_l \frac{j(j+1) + l(l+1) - s(s+1)}{2j(j+1)} + g_s \frac{j(j+1) - l(l+1) + s(s+1)}{2j(j+1)}$$

For $j = l + 1/2$, which holds for both nuclei since the valence nucleus is in $p_{3/2}$, the previous equation simplifies to

$$j g_j = g_l l + g_s/2$$

For ^{57}Cu, substituting the orbital and spin g-factors for a proton, $g_l = 1$, $g_s = +5.6$ n.m., we obtain

$$\mu(^{57}\text{Cu}) = j g_j = 1 \times 1 + 5.6/2 = 3.8 \text{ n.m.}$$

For ^{57}Ni, having a valence neutron, the g-factors are $g_l = 0$, $g_s = -3.8$ n.m. and we have

$$\mu(^{57}\text{Ni}) = j g_j = 0 \times 1 - 3.8/2 = -1.9 \text{ n.m.}$$

(c) Q_{β^+} is given by $[M(A, Z) - M(A, Z - 1) - 2m]c^2$, where M denotes the atomic mass and m the electron mass. Since the parent (^{57}Cu) and daughter (^{57}Ni) nuclei are *mirror* nuclei, the binding energy difference is only due to the difference in the Coulomb energies. To write the atomic mass difference we have only to subtract[3] the mass difference because a proton is exchanged into a neutron after the decay. Hence we can write

$$\Delta Mc^2 \approx E_c(Z) - E_c(Z - 1) + (M_p - M_n)c^2 \simeq$$

$$\simeq \frac{3}{5} \frac{e^2}{4\pi\epsilon_0 R} [Z^2 - (Z - 1)^2] + (M_p - M_n)c^2 \simeq \frac{3}{5} \frac{\alpha\hbar c}{r_0} \frac{2Z + 1}{A^{1/3}} + (M_p - M_n)c^2 \simeq$$

$$\simeq \frac{3}{5} \frac{197}{137 \times 1.2} \frac{59}{57^{1/3}} + 938.27 - 939.57 \simeq 9.72 \text{ MeV}$$

The maximum positron energy is equal to Q_{β^+} and then we have

$$T_{\max} = Q_{\beta^+} = (\Delta M - 2m)c^2 \approx 9.72 - 2 \times 0.511 \simeq 8.7 \text{ MeV.}$$

It is worth to notice that using the *SEMF* the result is 8.5 MeV.

Exercise 1.5.10

The shell sequence up to 14 is $1s_{1/2} \, 1p_{3/2} \, 1p_{1/2} \, 1d_{5/2}$.

The spin and parity of $^{17}_{8}$O is that of the uncomplete neutron shell:
- $^{17}_{8}$O, $n: (1d_{5/2})^1 \, l = 2, \, j = \frac{5}{2} \Rightarrow J^P = \frac{5}{2}^+$.

In the case of $^{18}_{9}$F, there are valence nucleons in both proton and neutron shells. Hence the shell model prediction is not unique. The valence shell is the same $1d_{5/2}$. The resulting spin comes from the angular momentum composition $\frac{5}{2} \oplus \frac{5}{2}$, whereas the parity is the product $(-1)^2 \times (-1)^2 = +1$. So we have
- $^{18}_{9}$F, $p: (1d_{5/2})^1$, n: $(1d_{5/2})^1 \Rightarrow J^P = 0^+, \, 1^+ \, 2^+, \, 3^+, \, 4^+, \, 5^+$.

(From measurements we have $J^P = 1^+$).

For the last nucleus we need to extend the shell sequence. The two last shells up to an occupation 82 are $2d_{3/2} \, 3s_{1/2}$, instead up to 126 are $1i_{13/2} \, 3p_{1/2}$. $^{207}_{82}$Pb$_{125}$ has a valence neutron in the $3p_{1/2}$ shell. Hence we find
- $^{207}_{82}$Pb, $n: (3p_{1/2})^1 \, l = 1, \, j = \frac{1}{2} \Rightarrow J^P = \frac{1}{2}^-$.

---

[3]In the *SEMF* the energy term and the mass term have opposite signs.

A.2 Solutions of Particle Physics (Chapter 2)

2.1 Fundamental Interactions

Exercise 2.1.1

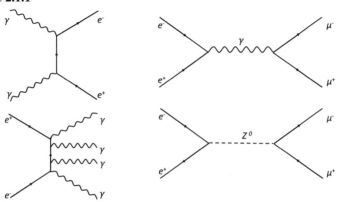

Exercise 2.1.2

Let us first convert the cross section from natural to CGS units. In natural units $G_F = 1.2 \times 10^{-5}$ GeV^{-2}; then expressing $\sqrt{s}$ in GeV, we obtain $G_F^2 s = 1.44 \times 10^{-10} \left(\frac{\sqrt{s}}{1\,\text{GeV}}\right)^2$ GeV^{-2}. To perform the conversion we use the relationship $\hbar c \simeq 197$ MeV·fm $\simeq 1.97 \times 10^{-14}$ GeV·cm, which allows to get $1/\text{GeV} = 1.97 \times 10^{-14}$ cm. Hence

$$G_F^2 s = 5.6 \times 10^{-38} \left(\frac{\sqrt{s}}{1\,\text{GeV}}\right)^2 \text{cm}^2.$$

The CMS square total energy of the ν-nucleon is given by the invariant $(p_\nu + p_p)^2$, where p_ν and p_p are the 4-momenta of the neutrino and proton respectively. Therefore $s = M_p^2 + M_\nu^2 + 2 M_p E_\nu$: substituting $M_\nu = 0$ and neglecting M_p^2 ($M_p = 0.94$ GeV/c^2) in the high energy limit, we get

$$s \simeq 2 M_p E_\nu \simeq 1.88 \left(\frac{E_\nu}{1\,\text{GeV}}\right) \text{GeV}^2$$

and for the cross section

$$\sigma \simeq \frac{2 \times 5.6\ 10^{-38} \times 1.88}{28.27} \left(\frac{E_\nu}{1\,\text{GeV}}\right) \simeq 7.4 \times 10^{-39} \left(\frac{E_\nu}{1\,\text{GeV}}\right) \text{cm}^2.$$

The number of scatterers (nucleons) per unit volume is $n = \rho/M_p = N_A \rho \simeq 1.3 \times 10^{24}$ cm^{-3} and hence the interaction length is

$$\lambda = \frac{1}{\sigma n} \simeq 1.0 \times 10^{14} \left(\frac{E_\nu}{1\,\text{GeV}} \right)^{-1} \text{cm.}$$

An estimate of the ν energy above which the Earth becomes opaque is obtained equating such length to the Earth diameter $D = 1.2 \times 10^9$ cm. This energy turns out to be $E_\nu > 8.3 \times 10^4$ GeV.

Exercise 2.1.3

- $e^+ + e^- \rightarrow \mu^+ + \mu^- : \gamma + Z^0$
- $n \rightarrow p + e^- + \bar{\nu}_e : W$
- $\mu^- \rightarrow e^- + \bar{\nu}_e + \nu_\mu : W$
- $\nu_e + e^- \rightarrow \nu_e + e^- : W + Z^0$
- $\nu_\mu + e^- \rightarrow \nu_\mu + e^- : Z^0$

Exercise 2.1.4

All the processes are allowed, except $p + p \rightarrow K^+ + p$, which is forbidden because of baryon conservation ($B_{\text{ini}} = 2 \neq B_{\text{fin}} = 1$) and strangeness conservation ($S_{\text{ini}} = 0 \neq S_{\text{fin}} = +1$). The first two processes

$$\gamma + \gamma \rightarrow \gamma + \gamma, \quad e^+ + e^- \rightarrow 4\gamma$$

are due to electromagnetic interaction, the third and the fifth ones

$$p + \bar{p} \rightarrow W^- + X, \quad \nu_\mu + e^- \rightarrow \nu_\mu + e^-$$

to weak interaction. The Feynman diagrams of the allowed reactions are shown below.

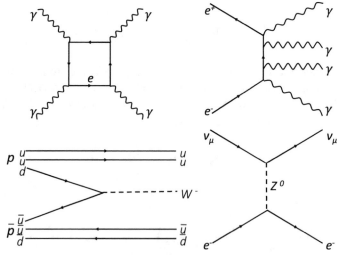

Exercise 2.1.5

- $e^+ + e^- \to \gamma + \gamma$: allowed—e.m. interaction
- $\pi^- + n \to K^- + \Lambda$: forbidden—strangeness not conserved ($K^- = s\bar{u}$, $\Lambda = uds$
  $\Rightarrow S_{\text{ini}} = 0 \neq S_{\text{fin}} = -2$).
- $\Sigma^+ \to n + e^+ + \nu_e$: forbidden—weak interaction but two flavors changed ($\Sigma^+ = uus$, $n = udd$).
- $\Sigma^+ \to \Lambda + e^+ + \nu_e$: allowed—weak interaction($u \to d + W^+$).
- $\rho^0 \to K^+ + K^-$: forbidden—kinematics ($m_{\text{fin}} > m_{\text{ini}}$).
- $\bar{\nu}_e + e^- \to \bar{\nu}_e + e^-$: allowed—weak interaction ($W^- + Z^0$)
- $\nu_e + e^- \to \nu_e + e^-$: allowed—weak interaction ($W^- + Z^0$).

The Feynman diagrams for the allowed processes are shown below.

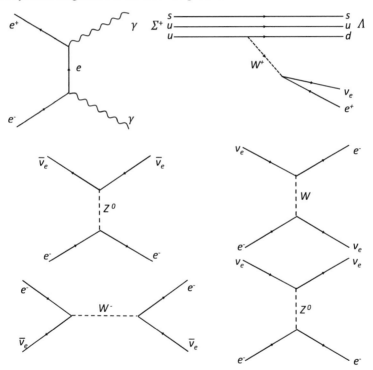

Exercise 2.1.6

a. $\pi^- + p \to \Sigma^0 + K^0$: it is a strong interaction process. X must have $Q = 0$, $B = 0$ and strangeness $S = +1$ ($\bar{s}$) (because $\Sigma^0 = uds$ and then $S = -1$). $K^0 = d\bar{s}$ possesses all these features.

b. $e^+ + n \to p + \bar{\nu}_e$: it is a weak process. X must have $Q = 0$ and electron lepton number $L_e = -1$. Hence it is an $\bar{\nu}_e$. The same result can be obtained using the neutron beta decay, $n \to p + e^- + \bar{\nu}_e$, and moving the electron to the initial state.

c. $\Xi^0 \rightarrow \Lambda + X$. The missing particle must be a meson. Looking at the particles involved (all hadrons) the decay can be either strong or weak.[4]

If the decay were strong, the meson should have $Q = 0$ and strangeness $S = -1$. It might be a $\bar{K}^0$, but the system $\Lambda + \bar{K}^0$ is too heavy for the Ξ^0 decay. So a strong decay is excluded.

The decay is weak and the strangeness conservation is not holding any more, being replaced by $\Delta S = \pm 1$: a neutral pion is the right answer. Hence it is $\Xi^0 \rightarrow \Lambda + \pi^0$.

Exercise 2.1.7

The figure below shows the Feynman diagrams for all the processes. It has to be noticed that (a) and (b) can also occur as neutral current processes (Z^0 instead of γ). This concurrence is more and more important as energy increases.

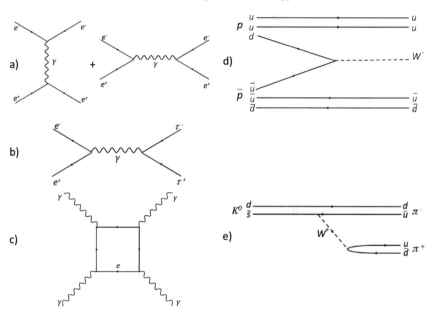

Exercise 2.1.8

Hereafter is the list of reactions (**A** for allowed, **F** for forbidden, "conservation" is implicit):

1. $\mu^+ \rightarrow e^+ + \gamma$: **F**, violates L_e and L_μ.
2. $e^- \rightarrow \nu_e + \gamma$: **F**, violates charge.
3. $p + p \rightarrow \Sigma^+ + K^+$: **F**, violates B.
4. $e^+ + e^- \rightarrow \gamma$: **F**, violates energy.
5. $\nu_\mu + p \rightarrow \mu^+ + n$: **F**, violates L_μ.
6. $\nu_\mu + n \rightarrow \mu^- + p$: **A**, see figure.

---

[4]For the moment we disregard the knowledge that all the $J^P = (1/2)^+$ baryons, except p and Σ^0, decay weakly.

7. $e^+ + n \rightarrow p + \nu_e$:   **F**, violates L_e

8. $e^- + p \rightarrow n + \nu_e$:   **A**, see figure.

9. $\pi^+ \rightarrow \pi^0 + e^+ + \nu_e$:   **A**, see figure for one of the possible graphs

10. $p + \bar{p} \rightarrow Z^0 + X$:   **A**, a possible case is shown, with $q\bar{q}$ fragmentations omitted.

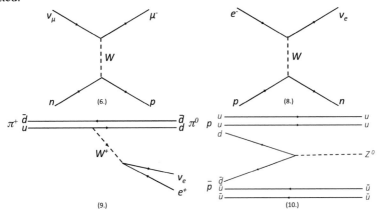

2.2   Hadrons

Exercise 2.2.1

(1) K^0-mesons as in (2.1) are produced in a strong interaction process. The following quantities are then conserved: electric charge Q, baryon number B, lepton number L, strangeness S. Considering the initial state and the K^0 in the final state, the requirements for X are $Q = +2$, $B = +2$, $L = 0$, $S = -1$. No known particle exists with such numbers. The minimum number of particles composing X is two because two baryons can realize a system with $B = +2$. Hereafter a few processes fulfilling these requirements are listed:

- $p + p \rightarrow K^0 + p + \Sigma^+$
- $p + p \rightarrow K^0 + \Delta^+ + \Sigma^+$
- $p + p \rightarrow K^0 + \Sigma^0 + \Delta^{++}$
- $p + p \rightarrow K^0 + \bar{K}^0 + p + p$
- $p + p \rightarrow K^0 + \bar{K}^0 + p + n + \pi^+$
- $p + p \rightarrow K^0 + K^0 + \Sigma^+ + \Sigma^+$
- $p + p \rightarrow K^0 + \Delta^0 + \Sigma^+ + \pi^+$
- $p + p \rightarrow K^0 + K^0 + \Sigma^0 + \Sigma^0 + \pi^+ + \pi^+$

We further notice that the minimum energy (threshold energy) is different for each of the listed processes.

(2) Several experimental set-ups can be used to study reaction (2.1), depending on the quantities to be measured and the particle identification required.

Let us assume that the experimental configuration consists in a beam of protons hitting a fixed target. Since we want to select events including K^0-mesons, the observation of their decays is mandatory. As K^0 is neutral, its decay is detected through the observation of the decay particles. One can use various types of detectors positioned downstream of the target (e.g., wire or drift chambers) or imaging detectors acting as target as well (e.g., bubble chamber). With such detectors it is possible to observe the decay into $\pi^+\pi^-$ from their tracks. To measure charge and momentum of the pions a suitable magnetic field is the best solution. The detection of the decay into neutral pions is much more challenging, because it requires the observation of the two photons emerging from the quasi immediate π^0 decay. This can be achieved with a downstream electromagnetic calorimeter or, in the case of a bubble chamber, using a heavy liquid filling (e.g. freon).

The K^0 decay follows the exponential decay law $N(t) = N_0 e^{-t/\tau}$, where $N(t)$ (N_0) is the number of particles at time t (time 0) and τ is the mean lifetime. It can be reasonable to require that 99% of the neutral kaons decay in the detector. This requirement determines the size of the experimental set-up. We have

$$0.99 = \int_0^T \frac{dt}{\tau} \frac{N(t)}{N_0} = 1 - e^{T/\tau}$$

hence $T \simeq 4.6\,\tau$. Therefore the minimum length of the experimental set-up is

$$L \simeq 4.6\,\beta\gamma c\tau = 4.6\,c\tau\frac{p}{m} \simeq 0.74 \text{ m}.$$

Exercise 2.2.2

All the reactions are strong interaction processes. Considering the particles involved we need to check the conservation of the following quantities: electric charge Q, baryon number B and strangeness S.

- $K^- + p \to \Omega^- + K^+ + K^0$: allowed;
- $\psi \to \pi^+ + \pi^0 + \pi^-$: allowed;
- $\pi^- + p \to \Sigma^+ + K^-$: forbidden for S non conservation;
- $\pi^- + p \to \pi^0 + \pi^0$: forbidden for B non conservation;
- $p + p \to n + \Delta^{++} + p + \bar{p}$: allowed.

Exercise 2.2.3

The answers about the decays and the interaction type are

- $\phi \to \rho^0 + \pi^0$: allowed, strong interaction;
- $\pi^0 \to e^+ + e^- + \gamma$: allowed, e.m. interaction;
- $\Xi^- \to \Sigma^0 + \mu^- + \bar{\nu}_e$: forbidden, violates the electron and muon numbers conservations;
- $\Sigma^- \to n + \pi^-$: allowed, weak interaction;
- $\Xi^- \to \pi^0 + \pi^-$: forbidden, violates the baryon number conservation.

Exercise 2.2.4

To get the $e^+ + e^- \rightarrow \mu^+ + \mu^-$ cross section in cm^2 we simply multiply it by $(\hbar c)^2$:

$$\sigma(\mu^+\mu^-) = \frac{4\pi\alpha^2}{3s}(\hbar c)^2 \simeq \frac{4\,3.14}{3s} \times \left(\frac{0.197\,\text{GeV} \cdot \text{fm}}{137}\right)^2 \simeq 86.6\,\text{nb}\,\left(\frac{\text{GeV}^2}{s}\right)$$

Neglecting strong interaction effects, the cross section into hadrons can be estimated from the ratio R

$$R = \frac{\sigma(\text{hadrons})}{\sigma(\mu^+\mu^-)} = C\sum_q Q_q^2$$

where C is the number of quark colors (3), Q_q is the charge of the quark q (in e units) and the sum includes those quarks for which $m(q\bar{q}) < \sqrt{s}$. At 2 GeV u, d e s fulfill such condition and then

$$\sigma(\text{hadrons}) = 3 \times \left(\frac{1}{9} + \frac{4}{9} + \frac{1}{9}\right) \times 86.6\,\text{nb}\,\left(\frac{\text{GeV}^2}{4\,\text{GeV}^2}\right) \simeq 43.3\,\text{nb}$$

Exercise 2.2.5

We have

$$\tau_{J/\psi} = \frac{\hbar}{\Gamma(J/\psi)} = \frac{\hbar c}{\Gamma(J/\psi)c} \simeq \frac{197\,\text{MeV fm}}{0.091\,\text{MeV}\,3\,10^{23}\,\text{fm/s}} \simeq 7.2 \times 10^{-21}\,\text{s}$$

The decay time corresponds to a strong interaction decay.

Exercise 2.2.6

The beam energy is above the energy threshold for the production of strange particles, but below that for producing particles with heavier quarks. Therefore the simplest hypothesis for the event is the associated production of Λ and K^0 observed through their respective decays into $p + \pi^-$ and $\pi^+ + \pi^-$. Having in mind also the two charged tracks, the simplest interpretation for the event is

$$\pi^+ + p \rightarrow \pi^+ + \pi^+ + \Lambda + K^0$$

To verify the correctness of the interpretation and to assign a specific particle to each V^0, we assume that the negative track is a π^-, whereas the positive one can be either p (Λ-hypothesis) or π^+ (K^0-hypothesis).

Let us call V_1^0 the first vertex. If it is a Λ decay, we have

$$M^2 = m_p^2 + m_\pi^2 + 2\sqrt{p_{1+}^2 + m_p^2}\sqrt{p_{1-}^2 + m_\pi^2} - 2p_{1+}p_{1-}\cos\theta_1 =$$

$$= 0.938^2 + 0.139^2 + 2 \times 1.02 \times 1.905 - 2 \times 0.4 \times 1.9 \times \cos 24.5° \simeq 3.40\,\text{GeV}^2$$

hence $M \simeq 1.84$ GeV, which is inconsistent with the hypothesis, since it differs by more than 5% from the Λ mass (1.116 GeV/c^2).

If V_1^0 is a K^0 decay, we have

$$M^2 = m_\pi^2 + m_\pi^2 + 2\sqrt{p_{1+}^2 + m_\pi^2}\sqrt{p_{1-}^2 + m_\pi^2} - 2p_{1+}p_{1-}\cos\theta_1 =$$

$$= 0.139^2 + 0.139^2 + 2 \times 0.423 \times 1.905 - 2 \times 0.4 \times 1.9 \times \cos 24.5° \simeq 0.267 \text{ GeV}^2$$

hence $M \simeq 0.517$ GeV, which is consistent with the hypothesis, being within 5% from the K^0 mass (0.498 GeV/c^2).

V_2^0 is the second vertex. If it is a Λ decay, we have

$$M^2 = m_p^2 + m_\pi^2 + 2\sqrt{p_{2+}^2 + m_p^2}\sqrt{p_{2-}^2 + m_\pi^2} - 2p_{2+}p_{2-}\cos\theta_2 =$$

$$= 0.938^2 + 0.139^2 + 2 \times 1.20 \times 0.286 - 2 \times 0.75 \times 0.25 \times \cos 22° \simeq 1.24 \text{ GeV}^2$$

hence $M \simeq 1.11$ GeV, which differs from the Λ mass by less than 5%.

To further confirm the Λ-hypothesis for V_2^0, we calculate the invariant mass for a K^0 as

$$M^2 = m_\pi^2 + m_\pi^2 + 2\sqrt{p_{2+}^2 + m_\pi^2}\sqrt{p_{2-}^2 + m_\pi^2} - 2p_{2+}p_{2-}\cos\theta_2 =$$

$$= 0.139^2 + 0.139^2 + 2 \times 0.76 \times 0.286 - 2 \times 0.75 \times 0.25 \times \cos 22° \simeq 0.126 \text{ GeV}^2$$

$M \simeq 0.354$ GeV is inconsistent with the K^0 mass.

As a conclusion V_1^0 is a K^0, V_2^0 is a Λ.

The lifetime of each particle is

$$t = \frac{l}{\beta\gamma c} = \frac{l}{c} \times \frac{m}{p}$$

where m and p are mass and momentum of the decaying particle. We have

$$p_{K^0} = \sqrt{p_{1+}^2 + p_{1-}^2 + 2p_{1+}p_{1-}\cos\theta_1} \simeq$$

$$\simeq \sqrt{0.4^2 + 1.9^2 + 2 \times 0.4 \times 1.9 \times \cos 24.5°} \simeq 2.27 \text{ GeV/c}$$

$$p_\Lambda = \sqrt{p_{2+}^2 + p_{2-}^2 + 2p_{2+}p_{2-}\cos\theta_2} \simeq$$

$$\simeq \sqrt{0.75^2 + 0.25^2 + 2 \times 0.75 \times 0.25 \times \cos 22°} \simeq 0.99 \text{ GeV/c}$$

and the lifetimes are

$$t_{K^0} \simeq \frac{37 \text{ cm}}{3 \times 10^{10} \text{ cm/s}} \times \frac{0.498}{2.27} \simeq 2.7 \ 10^{-10} \text{ s}$$

$$t_\Lambda \simeq \frac{11 \text{ cm}}{3 \times 10^{10} \text{ cm/s}} \times \frac{1.116}{0.99} \simeq 4.1 \ 10^{-10} \text{ s}$$

Exercise 2.2.7

a. Forbidden: strangeness is not conserved.
b. Forbidden: electric charge is not conserved.
c. Allowed.
d. Forbidden: energy is not conserved.
e. Forbidden: strangeness is not conserved.
f. Allowed.

Exercise 2.2.8

a. Σ^0 decays by electromagnetic interaction. For this interaction the quark flavor is conserved as for the strong interaction. The strangeness conserving decay is possible because a lighter baryon with the same strangeness does exist. The $\Sigma^0 \to \Lambda$ decay would be also possible by strong interaction if accompanied by π^0, but there is not enough energy $[M(\Sigma^0) < M(\Lambda) + M(\pi^0)]$. The electromagnetic decay is instead possible with the emission of a photon, which is kinematically allowed. The mean lifetime reflects the nature of the interaction.
b. Σ^+ cannot decay by strong interaction for the same reason as above $[M(\Sigma^+) < M(\Lambda) + M(\pi^+)]$. Nor can decay by e.m. interaction because there is no lighter charged baryon with $S = -1$. Hence it decays by weak interaction as shown by the mean lifetime.
c. Any diagram with a quark (among u, d and s) emitting a photon, because the quark contents of Σ^0 and Λ are the same and there is no flavor change.

Exercise 2.2.9

(a) Denoting by σ_Λ and p_Λ the Λ spin and momentum, and by p_K the K^0 momentum, the vector product $t = \sigma_\Lambda \times (p_\Lambda \times p_K)$ is parallel to the scattering plane and proportional to the Λ spin value. Hence it is proportional to the component of the spin in this plane. t is an axial vector and then must be zero if parity is conserved ($t \to -t$ under parity transformation). This is the case for the strong reaction $\pi^- + p \to \Lambda + K^0$. Being null the spin component in the scattering plane, the Λ spin can only be normal to this plane.
(b) Using the star superscript for center-of-momentum system (CMS) kinematic variables, we have the following relations

$$\epsilon_\pi = \sqrt{p_\pi^2 + m_\pi^2} = \sqrt{p_\pi^2 + m_\pi^2} \simeq 1.01 \text{ GeV/c}$$

$$E^* = \epsilon_\pi^* + \epsilon_p^* = \sqrt{m_\pi^2 + m_p^2 + 2m_p\epsilon_\pi} \simeq 1.67 \text{ GeV}$$

$$\beta_{CM} = |\boldsymbol{p}_\pi|/(\epsilon_\pi + \epsilon_p) = p_\pi/(\epsilon_\pi + m_p) \simeq 0.513$$

$$\gamma_{CM} = (\epsilon_\pi + \epsilon_p)/E^* \simeq 1.16$$

$$p^* = |\boldsymbol{p}^*| = \frac{\sqrt{[E^{*2} - (m_\Lambda + m_K)^2][E^{*2} - (m_\Lambda - m_K)^2]}}{2E^*} \simeq 0.203 \text{ GeV/c}$$

$$\epsilon_\Lambda^* = \sqrt{p^{*2} + m_\Lambda^2} \simeq 1.13 \text{ GeV.}$$

$\theta = 0$ in the Laboratory system (LS) corresponds to $\theta^* = 0$ in the CMS. Hence the Λ momentum in the LS is

$$p_\Lambda = \gamma_{CM} (p^* + \beta_{CM} \cdot \epsilon_\Lambda^*) \simeq 0.915 \text{ GeV/c}$$

The mean decay path of the Λ-particle is

$$\lambda_\Lambda = c\tau_\Lambda \cdot \beta_\Lambda \gamma_\Lambda = c\tau_\Lambda \cdot \frac{p_\Lambda}{m_\Lambda} \simeq 6.47 \text{ cm}$$

and the probability that it decays before reaching the detector is

$$P(< L) = \frac{1}{\lambda_\Lambda} \int_0^L \exp(-l/\lambda_\Lambda) \, dl \simeq 1 - \exp(-10/6.47) \simeq 79\%$$

(c) The precession angle at distance L from the target is

$$\phi = \omega t = \omega \frac{L}{v_\Lambda} = \omega \frac{L}{\beta_\Lambda c},$$

where ω is the Larmor angular frequency

$$\omega = \frac{\mu_\Lambda B}{\hbar}.$$

Hence we have

$$\phi = \frac{\mu_\Lambda B L}{\hbar c} \frac{E_\Lambda}{p_\Lambda} \simeq \frac{0.61 \times 3.15 \ 10^{-14} \text{ MeV/T} \times 20 \text{ T} \times 10 \text{ cm}}{197 \ 10^{-13} \text{ MeV cm}} \times \frac{\sqrt{0.915^2 + 1.116^2}}{0.915} \simeq$$

$$\simeq 0.308 \text{ rad} \simeq 17.6°$$

(d) The decay asymmetry is defined as

$$f_+ = \frac{\int_0^1 N(\cos\theta^*) \, d\cos\theta^*}{\int_{-1}^1 N(\cos\theta^*) \, d\cos\theta^*} = \frac{(x - \alpha x^2/2)|_0^1}{(x - \alpha x^2/2)|_{-1}^1} = \frac{1}{2}\left(1 - \frac{\alpha}{2}\right).$$

Thus we have for the asymmetry parameter α:

$$\alpha = 2(1 - 2f_+) \simeq 0.72$$

(e) The decay asymmetry is a consequence of the parity non conservation in weak interactions, as for the decay $\Lambda \to \pi^- + p$. In fact we have $N(\theta^*) \neq N(\pi - \theta^*)$.

Exercise 2.2.10

The event in the text is interpreted as $\pi^- + p \to \Lambda + K^0$. From momentum conservation, $p_\pi = p_\Lambda + p_K$, we derive the Λ momentum

$$p_\Lambda = p_\pi - p_K$$

whose absolute value is

$$p_\Lambda = \sqrt{p_\pi^2 + p_K^2 - 2|p_\pi||p_K|\cos\theta_K} \simeq$$

$$\simeq \sqrt{1.5^2 + 0.52^2 - 2 \times 1.5 \times 0.52 \times \cos 58°} \simeq 1.3\,\text{GeV/c}$$

a. Assuming that the particles decayed from the second V^0 are a proton (with momentum p_+) and a negative pion (with momentum p_-), the square of the invariant mass is

$$M^2 = m_p^2 + m_\pi^2 + 2\sqrt{p_+^2 + m_p^2}\sqrt{p_-^2 + m_\pi^2} - 2p_+ p_- \cos(\theta_+ + \theta_-) =$$

$$= 0.938^2 + 0.140^2 + 2 \times 1.31 \times 0.25 - 2 \times 0.92 \times 0.21 \times \cos 18° \simeq 1.195\,\text{GeV}^2$$

Hence the invariant mass is $\sqrt{1.195} \simeq 1.09$ GeV which does not correspond to a Λ-particle. The invariant mass is smaller: this implies that (at least) a neutral particle is not observed in the decay (as hypothesized in b.). This fact can be put in evidence using the momentum conservation in the longitudinal direction (i.e. along the Λ momentum). The total longitudinal momentum of the decay products is

$$p_+ \cos\theta_+ + p_- \cos\theta_- = 0.92\,\cos 4° + 0.21\,\cos 14° \simeq 1.12\,\text{GeV/c}$$

which is smaller than the Λ momentum (1.3 GeV/c) by more than $5\%\sqrt{2}$.

Finally we notice that the text did not provide the azimuthal angles of the decay products. Having these angles it would have been possible to evaluate the momentum vectors of these particles. The best way to verify the presence of unobserved neutrals is showing that p_Λ, p_+ and p_- are not lying in the same plane or equivalently the sum of the proton and pion transverse momenta is not zero.

b. If a neutrino is the missing neutral particle, its longitudinal momentum is

$$(p_\nu)_L = p_\Lambda - (p_+ \cos\theta_+ + p_- \cos\theta_-) \simeq 1.3 - 1.12 = 0.18\,\text{GeV/c}$$

c. The Λ lifetime is

$$t = \frac{l}{\beta \gamma c} = \frac{l}{c} \frac{m_\Lambda}{p_\Lambda} \simeq \frac{10 \text{ cm}}{3 \ 10^{10} \text{ cm/s}} \frac{1.116}{1.3} \simeq 2.04 \ 10^{-10} \text{ s.}$$

Exercise 2.2.11

(a) In the quark model baryons are 3-quark systems. Since quarks are fermions with spin 1/2, baryons must have a half-integer spin.

(b) An antibaryon is constituted of 3 antiquarks whose charges are either $-2/3$ or $+1/3$. Hence the maximum charge is $+1$ [$= 3 \times (+1/3)$].

(c) A meson is a quark-antiquark system. To get $S = -1$, the quark must be s whose charge is $Q_q = -1/3$. It follows that the charge of the meson can be either -1 ($Q_{\bar{q}} = -2/3$) or 0 ($Q_{\bar{q}} = +1/3$).

Exercise 2.2.12

(a) Mesons are $q\bar{q}$, the charges are $+2/3$ and $-1/3$ for q and $-2/3$ and $+1/3$ for $\bar{q}$. Combining the four possible cases, one finds that the charges for mesons are -1, 0 and $+1$.

(b) Antibaryons are $\bar{q}\bar{q}\bar{q}$. Again there are four possible cases which are $-2, -1, 0,$ $+1$.

2.3   Weak and Electro-Weak Interactions

Exercise 2.3.1

The neutrino mean free path in Iron is

$$\lambda = \frac{1}{n_p \sigma_\nu}$$

where $n_p = \rho_{Fe}/m_p$ is the number of nucleons per unit volume. We have

$$n_p \simeq \frac{7.9 \text{ g/cm}^3}{1.67 \ 10^{-24} \text{ g}} \simeq 4.7 \times 10^{24} \text{ cm}^{-3} \qquad \lambda = 7.1 \times 10^{10} \text{ cm}$$

Then, if $f = 1/10^9$ is the fraction of interacting neutrinos, the corresponding thickness is

$$L = f\lambda = 71 \text{ cm}$$

Exercise 2.3.2

For an estimate of the branching ratios we assume that they are simply proportional to the transition rates as given by the Fermi golden rule. Hence we have

$$\frac{BR(D^0 \to K^- e^+ \nu_e)}{BR(D^0 \to \pi^- e^+ \nu_e)} \simeq \frac{|M(D^0 \to K^- e^+ \nu_e)|^2}{|M(D^0 \to \pi^- e^+ \nu_e)|^2} \times \frac{\rho(D^0 \to K^- e^+ \nu_e)}{\rho(D^0 \to \pi^- e^+ \nu_e)},$$

where M denotes the transition amplitude and ρ the phase space factor. In the first ratio all the terms cancel but the effective coupling constants. These are $g_w \cos \theta_C$ for $D^0 \to K^- (c \to s + W^+)$ and $g_w \sin \theta_C$ for $D^0 \to \pi^-, (c \to d + W^+)$, where g_w is the weak coupling constant and θ_C is the Cabibbo angle ($\sin \theta_C \simeq 0.22$).

The phase space terms can be estimated using the so called *Sargent rule*, originally established for the beta decay, taking into account the kinematic analogy of the present decays with the beta case. Following this rule we have $w \propto E_0^5$, where w is the transition rate and E_0 is the energy available in the decay ($= m_n - m_p - m_e$, for the beta decay $n \to p + e^- + \bar{\nu}_e$). We also recall that in the Fermi theory the beta decay is only 'kinematical', that means that the energy dependence is only due to phase space. Therefore for the phase space we can write $\rho \propto E_0^5$. Under this assumption we have

$$\frac{BR(D^0 \to K^- e^+ \nu_e)}{BR(D^0 \to \pi^- e^+ \nu_e)} = \frac{\cos^2 \theta_C}{\sin^2 \theta_C} \times \left(\frac{m_D - m_K - m_e}{m_D - m_\pi - m_e} \right)^5 \simeq 20 \times 0.32 \simeq 6.4$$

Despite the crudeness of the estimate, this results differs from the experimental value by only 40%.

Exercise 2.3.3

The Feynman diagrams are reported below

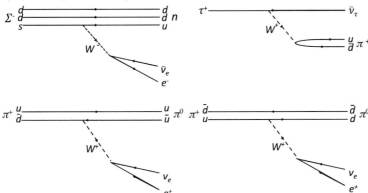

Exercise 2.3.4

◇ The beta decay rate in the limit of the *Sargent rule*, i.e. assuming $E \gg mc^2$ and substituting E_0 with $T_{\max} (\simeq 0.782 \text{ MeV})$ is

$$\omega = \frac{G_F^2}{2\pi^3 \hbar^7 c^6} \frac{T_{\max}^5}{30}. \tag{2.1}$$

The squared Fermi constant (divided by $(\hbar c)^3$, as it is usually expressed) is

$$\left[\frac{G_F}{(\hbar c)^3} \right]^2 = \frac{\omega \, 2\pi^3 \, (\hbar c) \, 30}{c \, T_{\max}^5} = \frac{\frac{1}{886 \text{ s}} \times 62 \times 197 \text{ MeV fm} \times 30}{3 \cdot 10^{23} \text{ fm s}^{-1} \times (0.782 \text{ MeV})^5} \simeq 4.7 \, 10^{-21} \text{ MeV}^{-4},$$

hence we have

$$\frac{G_F}{(\hbar c)^3} \simeq 6.9 \; 10^{-11} \; \text{MeV}^{-2} = 6.9 \; 10^{-5} \; \text{GeV}^{-2}.$$

The value is different from the one reported in the literature $(1.17 \; 10^{-5} \; \text{GeV}^{-2})$ because of the spectrum integration inaccuracy implicit in the *Sargent rule* and other aspects of Fermi theory not included in Eq. (2.1), e.g. the $V - A$ feature of weak interaction and the quark structure of the neutron.

⋄ Using the *Sargent rule* we have for the $^{35}_{16}S \rightarrow ^{35}_{17}Cl + e^- + \bar{\nu}_e$ decay

$$\frac{\omega[^{35}S]}{\omega[n]} = \left(\frac{Q[^{35}S]}{Q[n]}\right)^5 = \left(\frac{0.168}{0.782}\right)^5 \simeq 0.00046,$$

and then

$$\tau[^{35}S] = \frac{886 \; s}{0.00046} \simeq 1.9 \; 10^6 \; s \simeq 22 \; d.$$

⋄ Both the parent and daughter nuclei are odd-A, hence the spin-parity is determined by the unpaired nucleon. The shell occupation of this nucleon is

- $^{35}_{16}S \quad n : (1s_{1/2})^2(1p_{3/2})^4(1p_{1/2})^2(1d_{5/2})^6(2s_{1/2})^2(1d_{3/2})^3$
- $^{35}_{17}Cl \quad p : (1s_{1/2})^2(1p_{3/2})^4(1p_{1/2})^2(1d_{5/2})^6(2s_{1/2})^2(1d_{3/2})^1$

The unpaired nucleons have $l = 2$: both $^{35}_{16}S$ and $^{35}_{17}Cl$ have $J = 3/2$, $P = (-1)^2 = +$ $\Longrightarrow J^P = 3/2^+$.

Exercise 2.3.5

(1) Charged current ν_μ-interactions on nucleon valence quarks can be either

$$\nu_\mu + d \rightarrow \mu^- + u$$

or the ones associated to charm production

$$\nu_\mu + d \rightarrow \mu^- + c.$$

In these processes, the leptonic vertex is the same whereas the hadronic one is $g_W \cos\theta_C$ in the former and $g_W \sin\theta_C$ in the latter case, where θ_C is the Cabibbo angle ($\sin\theta_C \simeq 0.22$). The fraction of charm events in CC interactions can be estimated as

$$\frac{\sigma(\nu_\mu + d \rightarrow \mu^- + c)}{\sigma(\nu_\mu + d \rightarrow \mu^- + u) + \sigma(\nu_\mu + d \rightarrow \mu^- + c)} = \frac{\sin^2\theta_C}{\cos^2\theta_C + \sin^2\theta_C} =$$

$$= \sin^2\theta_C \simeq 0.05$$

(2) The probability for muon neutrinos to be detected as tau neutrinos is $P_{\mu\tau}$ whereas $1 - P_{\mu\tau}$ is the probability to survive in the initial state. The signal-to-noise ratio is then

$$r = \frac{N(\nu_\tau \to \tau^-)}{N(\nu_\mu + d \to \mu^- + c)} \simeq \frac{P_{\mu\tau}}{1 - P_{\mu\tau}} \times \frac{1}{\sin^2 \theta_C} \simeq \frac{0.015}{0.985} \times \frac{1}{0.22^2} \simeq 0.31.$$

(3) The τ^- decay modes are of type $\tau^- \to W^- + \nu_\tau$. A few cases are given below

- $\tau^- \to \mu^- + \bar{\nu}_\mu + \nu_\tau$ $[W^- \to \mu^- + \bar{\nu}_\mu]$
- $\tau^- \to e^- + \bar{\nu}_e + \nu_\tau$ $[W^- \to e^- + \bar{\nu}_e]$
- $\tau^- \to \pi^- + \nu_\tau$ $[W^- \to d + \bar{u}]$
- etc.

The Feynman graphs for these decays are shown below.

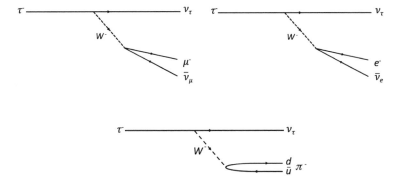

Exercise 2.3.6

Following the Fermi golden rule, the branching ratio is proportional to the absolute square of the transition amplitude times the phase space factor. In the decays of the text we have

$$\frac{BR(\Sigma^- \to n + e^- + \bar{\nu}_e)}{BR(\Sigma^- \to \Lambda + e^- + \bar{\nu}_e)} \simeq \frac{|M(\Sigma^- \to n + e^- + \bar{\nu}_e)|^2}{|M(\Sigma^- \to \Lambda + e^- + \bar{\nu}_e)|^2} \times \frac{\rho(\Sigma^- \to n + e^- + \bar{\nu}_e)}{\rho(\Sigma^- \to \Lambda + e^- + \bar{\nu}_e)}$$

For $\Sigma^- \to n$ we have an effective coupling constant $g_w \sin \theta_C$ ($dds \to ddu$ involves $s \to u + W^-$), with θ_C the Cabibbo angle. Instead for $\Sigma^- \to \Lambda$ we have $g_w \cos \theta_C$ ($dds \to uds$ involves $d \to u + W^-$).

The phase space factors (ρ) can be estimated using the *Sargent rule*. It is written as $w \propto E_0^5$, where w is the decay rate and E_0 is the energy available in the decay

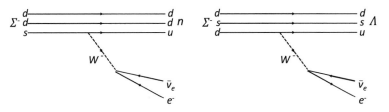

$(= m_n - m_p - m_e$, in the case of the neutron decay $n \to p + e^- + \bar{\nu}_e)$. In the Fermi theory of beta decay the transition rate is entirely due to kinematics. Therefore we can use the same expression for the phase space factor. So we have

$$\frac{BR(\Sigma^- \to n + e^- + \bar{\nu}_e)}{BR(\Sigma^- \to \Lambda + e^- + \bar{\nu}_e)} \simeq \frac{\sin^2 \theta_C}{\cos^2 \theta_C} \times \left(\frac{m_\Sigma - m_n - m_e}{m_\Sigma - m_\Lambda - m_e} \right)^5$$

and then

$$\tan^2 \theta_C \simeq \frac{BR(\Sigma^- \to n + e^- + \bar{\nu}_e)}{BR(\Sigma^- \to \Lambda + e^- + \bar{\nu}_e)} \times \left(\frac{m_\Sigma - m_\Lambda - m_e}{m_\Sigma - m_n - m_e} \right)^5 \simeq$$

$$\simeq \frac{10.2}{0.57} \times \left(\frac{1197 - 1116}{1197 - 940} \right)^5 \simeq 0.056$$

Hence $\sin\theta_C \simeq 0.23$, which is in good agreement with the known value ($\sin\theta_C \simeq 0.22$).

Exercise 2.3.7

Considering the transition amplitudes, for $D^+ \to \bar{K}^0 + e^+ + \nu_e$ we have $c \to s + W^+$ and an effective coupling constant $g_W \cos \theta_C$; for $\mu^+ \to e^+ + \nu_e + \bar{\nu}_\mu$ we have a pure leptonic vertex $\mu^+ \to \bar{\nu}_\mu + W^+$ and thus only g_W. Making use of the *Sargent rule* for the phase space factors we get

$$\frac{\Gamma(D^+ \to \bar{K}^0 + e^+ + \nu_e)}{\Gamma(\mu^+ \to e^+ + \nu_e + \bar{\nu}_\mu)} = \frac{|M(D^+ \to \bar{K}^0 + e^+ + \nu_e)|^2}{|M(\mu^+ \to e^+ + \nu_e + \bar{\nu}_\mu))|^2} \times \frac{\rho(D^+ \to \bar{K}^0 + e^+ + \nu_e)}{\rho(\mu^+ \to e^+ + \nu_e + \bar{\nu}_\mu)}$$

$$= \cos^2 \theta_C \times \left(\frac{m_{D^+} - m_{\bar{K}^0} - m_e}{m_\mu - m_e} \right)^5 \simeq 0.95^2 \times \left(\frac{1870 - 498 - 0.5}{106 - 0.5} \right)^5 \simeq 3.5 \times 10^5$$

The experimental value is 1.5×10^5.

Exercise 2.3.8

Pions produced in the atmospheric showers decay as $\pi^- \to \mu^- + \bar{\nu}_\mu$ and $\pi^+ \to \mu^+ + \nu_\mu$. The muons produced in this way decay as $\mu^- \to e^- + \bar{\nu}_e + \nu_\mu$ and $\mu^+ \to e^+ + \nu_e + \bar{\nu}_\mu$. All the charges have the same probability.

The pions produced in the hadronic interactions with atmosphere nuclei have energies higher than the ones observed for the atmospheric neutrinos. Let us assume that the pion energy is at most 1 GeV. The pion mean free path is

$$l_\pi = \beta\gamma c\tau_\pi = \frac{p_\pi}{m_\pi} c\tau_\pi < \frac{1}{0.140} 3 \ 10^8 \times 2.6 \ 10^{-8} \ \text{m} \simeq 55 \ \text{m}.$$

Since their production height is around 10 km, all the pions decay before reaching the ground, unless they interact with the atmosphere again. Under the same assumption, the produced muons have a mean free path

$$l_\mu = \beta \gamma c \tau_\mu = \frac{p_\mu}{m_\mu} c \tau_\mu < \frac{1}{0.106} \, 3 \, 10^5 \times 2.2 \, 10^{-6} \, \text{km} \simeq 6 \, \text{km}$$

and also muons preferentially decay. Counting all the types of neutrinos appearing in the decays we obtain a flavor ratio

$$\frac{\nu_\mu + \bar{\nu}_\mu}{\nu_e + \bar{\nu}_e} \simeq 2$$

Exercise 2.3.9

The Feynman graphs for $\mu^- \to e^- + \bar{\nu}_e + \nu_\mu$ and $\tau^- \to e^- + \bar{\nu}_e + \nu_\tau$ are identical apart the masses involved. Recalling the *Sargent rule*, one gets:

$$R = \frac{\Gamma(\tau^- \to e^- + \bar{\nu}_e + \nu_\tau)}{\Gamma(\mu^- \to e^- + \bar{\nu}_e + \nu_\mu)} = \left(\frac{m_\tau - m_e}{m_\mu - m_e} \right)^5 \simeq \left(\frac{m_\tau}{m_\mu} \right)^5 \simeq 1.32 \times 10^6$$

Denoting with $B(\tau^- \to e^- + \bar{\nu}_e + \nu_\tau)$ the branching ratio of this mode, the tau mean lifetime is then

$$\tau_\tau = \frac{\tau_\mu}{R} \times B(\tau^- \to e^- + \bar{\nu}_e + \nu_\tau) \simeq \frac{2.2 \times 10^{-6}}{1.32 \times 10^6} \times 0.18 \simeq 3 \times 10^{-13} \, \text{s}$$

A.3 Solutions of Experiments and Detection Methods (Chapter 3)

3.1 Kinematics

Exercise 3.1.1

Here we write a few kinematical relations useful for the solution:

$$\epsilon_\pi = \sqrt{p_\pi^2 + m_\pi^2} = \sqrt{p_\pi^2 + m_\pi^2} \simeq 20 \, \text{GeV}$$
$$E^* = \epsilon_\pi^* + \epsilon_p^* = \sqrt{m_\pi^2 + m_p^2 + 2 m_p \epsilon_\pi} \simeq 6.199 \, \text{GeV}$$
$$\beta_{CM} = |\boldsymbol{p}_\pi|/(\epsilon_\pi + \epsilon_p) = p_\pi/(\epsilon_\pi + m_p) \simeq 0.955165$$
$$\gamma_{CM} = (\epsilon_\pi + \epsilon_p)/E^* \simeq 3.3775$$
$$p^* = |\boldsymbol{p}^*| = \frac{\sqrt{[E^{*2} - (m_\Sigma + m_K)^2][E^{*2} - (m_\Sigma - m_K)^2]}}{2 E^*} \simeq 2.965 \, \text{GeV}/c$$

1. Neglecting the thickness of the target, detectable tracks are produced by ionising particles emitted between $0°$ and $90°$ in the Laboratory system (LS). To answer the first question we have to establish if Σ^+'s produced in the experiment do exhibit a maximum angle. We have

$$\epsilon_\Sigma^* = \sqrt{p^{*2} + m_\Sigma^2} \simeq 3.194 \, \text{GeV}$$

$$\beta_\Sigma^* = p^*/\epsilon_\Sigma^* \simeq 0.9281$$

$\beta_\Sigma^* < \beta_{CM}$ is the condition to have such limiting angle and hence all Σ^+'s can be detected.[5]

2. Assuming that all Σ^+'s decay within 3 mean lifetimes, the maximum distance for the decay point is

$$D_\Sigma = 3 \cdot c\tau_\Sigma \cdot \beta_\Sigma \gamma_\Sigma = 3 \cdot c\tau_\Sigma \cdot \frac{p_\Sigma}{m_\Sigma} \simeq 6.05 \times \frac{p_\Sigma}{\text{GeV}/c} \text{ cm},$$

along the direction $\boldsymbol{p}_\Sigma/p_\Sigma$. From this expression we desume that the minimum length for the tracker corresponds to the maximum longitudinal momentum $(p_\Sigma)_L$. This occurs for $(p^*\Sigma)_L = p^*$. Hence we have

$$(p_\Sigma)_L^{max} = \gamma_{CM}(p^* + \beta_{CM} \cdot \epsilon_\Sigma^*) \simeq 20.3 \text{ GeV}/c \implies$$

$$L = 6.05 \times 20.3 \text{ cm} \simeq 122.8 \text{ cm}$$

3. As in the previous case, the minimum radius corresponds to the maximum $(p_\Sigma)_T$, that is for $(p^*\Sigma)_T = p^*$

$$R = 6.05 \times \frac{(p_\Sigma)_T^{max}}{\text{GeV}/c} = 6.05 \times 2.965 \text{ cm} \simeq 17.9 \text{ cm}$$

4. To establish if there is a maximum angle for K^+, we calculate its velocity in the CMS

$$\epsilon_K^* = \sqrt{p^{*2} + m_K^2} \simeq 3.006 \text{ GeV}$$

$$\beta_K^* = p^*/\epsilon_K^* \simeq 0.9864$$

We have $\beta_K^* > \beta_{CM}$, so there is no limiting angle. Hence kaons can escape from the tracker.

---

[5]It can be useful, though not necessary, to calculate the maximum angle and its corresponding angle in the CMS. They are

$$\tan\theta_{max} = \beta_\Sigma^* / (\gamma_{CM}\sqrt{\beta_{CM}^2 - \beta_\Sigma^{*2}}) \simeq 1.217 \implies \theta_{max} \simeq 50.6°$$

$$\cos\theta^*(\theta_{max}) = -\beta_\Sigma^* / \beta_{CM} \simeq -0.9717 \implies \theta^*(\theta_{max}) \simeq 166°$$

At the maximum angle the longitudinal, transverse and total Σ^+-momenta are respectively

$$(p_\Sigma)_L = \gamma_{CM}(p^* \cdot \cos\theta^*(\theta_{max}) + \beta_{CM} \cdot \epsilon_\Sigma^*) \simeq 0.573 \text{ GeV}/c$$

$$(p_\Sigma)_T = (p_\Sigma^*)_T = p^* \cdot \sin\theta^*(\theta_{max}) \simeq 0.700 \text{ GeV}/c$$

$$p_\Sigma \simeq 0.905 \text{ GeV}/c.$$

5. The detectable kaons are those produced in the forward direction in the LS ($0° < \theta < 90°$). The CMS angle corresponding to $\theta = 90°$ can be obtained from the Lorentz transformation of the kaon longitudinal momentum $(p_K)_L = \gamma_{CM} [(p_K^*)_L + \beta_{CM} \cdot \epsilon_K^*]$ by setting $(p_K)_L = 0$. Hence we have

$$(p_K^*)_L = p^* \cos \theta^*(90°) = -\beta_{CM} \, \epsilon_K^* \implies$$

$$\cos \theta^*(90°) = -\beta_{CM} \cdot \frac{\epsilon_K^*}{p^*} = -\frac{\beta_{CM}}{\beta_K^*} \simeq -0.9683,$$

corresponding to angle of about $165.5°$.

In the CMS frame the kaon angular distribution is isotropic so it is given by $dN/d\Omega = 1/4\pi$ (normalized to unity). The fraction of detectable kaons is then

$$r = \frac{1}{4\pi} \int_0^{2\pi} d\phi \int_{\theta^*(90°)}^1 d\cos\theta^* = \frac{1 - \cos\theta^*(90°)}{2} \simeq \frac{1.9683}{2} \simeq 98.4\%$$

Exercise 3.1.2

(1) To have a limiting production angle, particle 1 must fulfill the condition $\beta_{CM} \geq \beta_1^*$. The maximum limiting angle, corresponding to $90°$, is obtained for the equality in the previous relation. The CMS energy is the mass of the resonance, $E^* = M$. Hence we have

$$p^* = |\mathbf{p}^*| = \sqrt{[M^2 - (m_1 + m_2)^2][M^2 - (m_1 - m_2)^2]} \, / \, 2M$$

$$\epsilon_1^* = (M^2 + m_1^2 - m_2^2) \, / \, 2M$$

and then

$$\beta_1^* = \frac{p^*}{\epsilon_1^*} = \frac{\sqrt{[M^2 - (m_1 + m_2)^2][M^2 - (m_1 - m_2)^2]}}{M^2 + m_1^2 - m_2^2}$$

Since m_2 is negligible with respect to m_1, we get

$$\beta_1^* = \frac{M^2 - m_1^2}{M^2 + m_1^2} = \frac{2.58^2 - 1}{2.58^2 + 1} \simeq 0.7388$$

We can get the pion beam energy E_π, solving the equation $\beta_{CM} = \beta_1^*$

$$\beta_{CM} = \frac{|\mathbf{p}_\pi|}{E_\pi + m_p} = \frac{\sqrt{E_\pi^2 - m_\pi^2}}{E_\pi + m_p} = \beta_1^*.$$

Solving it in E_π we have

$$E_\pi = \frac{\beta_1^{*2} m_p + \sqrt{(\beta_1^{*2} m_p)^2 + (1 - \beta_1^{*2})(m_\pi^2 + \beta_1^{*2} m_p^2)}}{1 - \beta_1^{*2}} \simeq 2.65 \text{ GeV}.$$

For higher E_π values particle 1 is produced up to angles less than $90°$.
(2) Considering the decay $\Delta(2420) \rightarrow \Sigma + K$, for a fixed CMS angle $\theta^* = 120°$ and $\beta_{CM} = 0.7388$, we have

$$p^* = |p^*| = \frac{\sqrt{[M^2 - (m_\Sigma + m_K)^2][M^2 - (m_\Sigma - m_K)^2]}}{2M} \simeq 0.833 \text{ GeV/c}$$

$$\epsilon_\Sigma^* = \sqrt{p^{*2} + m_\Sigma^2} \simeq 1.452 \text{ GeV}$$

$$(p_\Sigma)_L = \gamma_{CM} (p^* \cos\theta^* + \beta_{CM} \cdot \epsilon_\Sigma^*) \simeq 0.974 \text{ GeV/c}$$

$$(p_\Sigma)_T = p^* \sin\theta^* \simeq 0.721 \text{ GeV/c}.$$

From the last expressions we can get the momentum and angle of the Σ in the Laboratory system

$$p_\Sigma = \sqrt{(p_\Sigma)_L^2 + (p_\Sigma)_T^2} \simeq 1.21 \text{ GeV/c}$$

$$\theta_\Sigma = \arccos \frac{(p_\Sigma)_L}{p_\Sigma} \simeq 36.5°$$

(3) The Σ-decay mean pathlength is $c\tau_\Sigma \cdot \beta_\Sigma \gamma_\Sigma = c\tau_\Sigma \cdot p_\Sigma/m_\Sigma$. The length of the detector is determined by the Σ's decaying in the forward direction, for which the momentum is maximum. This is

$$p_\Sigma^{\max} = \gamma_{CM} (p^* + \beta_{CM} \cdot \epsilon_\Sigma^*) \simeq 2.83 \text{ GeV/c}$$

The designed length corresponds to the requirement that 99% of the decay points are contained in the detector. This occurs for a proper time T so that we have

$$0.99 = \int_0^T \frac{1}{\tau_\Sigma N_0} N(t) dt = 1 - \int_T^\infty \frac{1}{\tau_\Sigma} \exp\left(-\frac{t}{\tau_\Sigma}\right) dt = 1 - \exp\left(-\frac{T}{\tau_\Sigma}\right)$$

$$\implies \quad T = -\ln(0.01) \cdot \tau_\Sigma \simeq 4.6 \cdot \tau_\Sigma$$

Hence the length of the detector must be

$$L = cT \cdot \frac{p_\Sigma^{\max}}{m_\Sigma} \simeq 4.6 \, c \, \tau_\Sigma \cdot \frac{p_\Sigma^{\max}}{m_\Sigma}$$

Solving this equation in τ_Σ, we finally get

$$\tau_\Sigma = \frac{L\, m_\Sigma}{4.6\, c\, p_\Sigma^{\max}} \simeq \frac{0.26 \times 1.189}{4.6 \times 3 \cdot 10^8 \times 2.83} \simeq 0.79 \times 10^{-10}\ \text{s}$$

Exercise 3.1.3

1. The minimum energy for a reaction is its threshold energy (E_{th}). It corresponds to the production of the final particles at rest in the CMS. Equating the 4-momentum invariants in the LS for the initial state and in the CMS for the final state, we have

$$M_\pi^2 + M_p^2 + 2 E_{\text{th}} M_p = (M_\Lambda + M_K)^2$$

and then

$$E_{\text{th}} = \frac{(M_\Lambda + M_K)^2 - M_\pi^2 - M_p^2}{2 M_p} \simeq 0.91\ \text{GeV}$$

2. A maximum production angle is possible provided that $\beta^* < \beta_{\text{CM}}$, where β^* is the CMS velocity of the particle and β_{CM} is the velocity of the CMS with respect to the LS. For $E_\pi = 2$ GeV, we have

$$\beta_{\text{CM}} = \frac{p_\pi}{E_\pi + M_p} \simeq 0.68$$

To get the Λ velocity in the CMS, we first calculate the total CMS energy (P_π and P_p are the 4-momenta of the pion and proton respectively)

$$E^* = \sqrt{(P_\pi + P_p)^2} = \sqrt{M_p^2 + M_\pi^2 + 2 E_\pi M_p} \simeq 2.16\ \text{GeV}$$

The momentum in the CMS is

$$p^* = |\mathbf{p}^*| = \frac{\sqrt{[E^{*2} - (M_\Lambda + M_K)^2][E^{*2} - (M_\Lambda - M_K)^2]}}{2\, E^*} \simeq 0.69\ \text{GeV}/c$$

Hence we have for the Λ-velocity in the CMS

$$\beta^* = \frac{p^*}{\sqrt{p^{*2} + M_\Lambda^2}} \simeq 0.52.$$

The condition $\beta^* < \beta_{\text{CM}}$ is fulfilled so that there is a maximum production angle for the Λ's. This angle turns out to be

$$\theta_{\max} = \arctan\left\{\left[\gamma_{\text{CM}}\sqrt{\left(\frac{\beta_{\text{CM}}}{\beta^*}\right)^2 - 1}\right]^{-1}\right\} \simeq 0.73\ \text{rad} \simeq 42°$$

Exercise 3.1.4

The invariant mass of the two-pion system is the mass M_X of the observed neutral particle

$$M_X^2 = (P_{\pi^+} + P_{\pi^-})^2 = 2M_\pi^2 + 2E_{\pi^+}E_{\pi^-} - 2p_{\pi^+}p_{\pi^-}\cos\theta.$$

The minimum opening angle corresponds to the case in which the two pions have the same energy $E_{\pi^+} = E_{\pi^-}(= E_X/2)$. Imposing this condition and having in mind that $E_\pi \gg M_\pi$ we get

$$M_X = \sqrt{E_X^2\sin^2\theta/2 + 2\,M_\pi^2} \simeq 0.495\ \text{GeV}/c^2$$

Exercise 3.1.5

Denoting by P_i the 4-momentum of electron i, the total CMS energy is written as

$$E^* = \sqrt{(P_1 + P_2)^2} = \sqrt{(E_1 + E_2)^2 - (p_1 + p_2)^2} =$$

$$\sqrt{E_1^2 + E_2^2 + 2E_1E_2 - p_1^2 - p_2^2 + 2p_1p_2} = \sqrt{2m^2 + 2E_1E_2 + 2p_1p_2} \simeq \sqrt{4E_1E_2}$$

where, in the last step, we have neglected the electron masses with respect to their energies. Therefore we have $E^* = 15.5\ \text{GeV}$.

In the CMS the two electron momenta are opposite. Neglecting the masses we have

$$p^* = \frac{E^*}{2} \simeq 7.74\ \text{GeV}/c.$$

The CMS velocity (in c units) in the LS is given by

$$\beta_{\text{CM}} = \frac{|p_1 + p_2|}{E_1 + E_2} = \frac{\sqrt{E_1^2 - m^2} - \sqrt{E_2^2 - m^2}}{E_1 + E_2} \simeq \frac{E_1 - E_2}{E_1 + E_2} \simeq 0.4 \qquad (3.1)$$

and the Lorentz factor is

$$\gamma_{\text{CM}} = (1 - \beta_{\text{CM}}^2)^{-1/2} \simeq 1.1.$$

If $E_1 = E_2$ and $p_1 = -p_2$, from Eq. (3.1) we get $\beta_{\text{CM}} = 0$. Hence the center-of-momentum and laboratory systems are coincident.

Exercise 3.1.6

The minimum electron energy is its rest mass ($\simeq 0.511\ \text{MeV}$), corresponding to the emission of an electron at rest.

To evaluate the maximum energy in a three-body decay $M \to m_1 + m_2 + m_3$, it is convenient to re-write it as a two-body decay $M \to M_{12} + m_3$, where M_{12} is the invariant mass of particles 1 and 2. It then turns out that the maximum energy for 3 is obtained when M_{12} is minimum, that is when it is equal to the $m_1 + m_2$. Hence we have

$$(E_3)_{\max} = \frac{M^2 + m_3^2 - (m_1 + m_2)^2}{2M}.$$

In our case $M = M_{\Xi^0}$, $m_3 = M_{e^-}$, $m_1 = M_{\Sigma^+}$ and $m_2 = M_\nu = 0$ and we have

$$(E_{e^-})_{\max} \simeq \frac{1315^2 + 0.511^2 - 1189^2}{2 \times 1315} \simeq 120 \text{ MeV}$$

Exercise 3.1.7

The minimum opening angle for a decay into two equal (ultra-relativistic) particles is obtained for

$$E_{\pi^+} = E_{\pi^-} = \frac{E_D}{2}.$$

Hence for the minimum opening angle between the pions we have

$$\theta_{\min} = \arcsin\left(\frac{\sqrt{M_D^2 - 2M_\pi^2}}{E_\pi}\right) = \arcsin\left(2\frac{\sqrt{M_D^2 - 2M_\pi^2}}{E_D}\right) \qquad (3.2)$$

To get the $\bar{D}^0$ energy, we make use of the knowledge that this particle is produced at the maximum angle $\theta_{\max}$. The corresponding angle in the CMS is given by the equation

$$\cos\bar{\theta}^* = \cos\theta^*(\theta_{\max}) = -\frac{\beta_D^*}{\beta_{\text{CM}}},$$

where β_D^* is the $\bar{D}^0$ velocity in the CMS and β_{CM} is the CMS velocity in the LS. Denoting by E_D^* and p^* the energy and momentum of the $\bar{D}^0$-particle in the CMS, using the Lorentz transformation for the energy, we get

$$E_D = \gamma_{\text{CM}}(E_D^* + \beta_{\text{CM}} p^* \cos\bar{\theta}^*) = \gamma_{\text{CM}}(E_D^* - \beta_D^* p^*). \qquad (3.3)$$

The total energy in the CMS is

$$E^* = \sqrt{(P_\pi + P_p)^2} = \sqrt{2E_\pi M_p + M_\pi^2 + M_p^2} = 6.21 \text{ GeV}.$$

Hence the $\bar{D}^0$-momentum in the CMS is

$$p^* = \frac{\sqrt{[E^{*2} - (M_\Sigma - M_D)^2][E^{*2} - (M_\Sigma + M_D)^2]}}{2E^*} \simeq 2.22 \text{ GeV}/c$$

and the corresponding energy is $\sqrt{p^{*2} + M_D^2} \simeq 2.90$ GeV.

For the quantities appearing in Eq. (3.3) we get

$$\beta_D^* = \frac{p^*}{\sqrt{p^{*2} + M_D^2}} = 0.767 \qquad \gamma_{CM} = \frac{E_\pi + M_p}{E^*} = 3.37,$$

and hence we have $E_D = 4.0$ GeV. Using (3.2), we finally get for the minimum opening angle

$$\theta_{\min} = \arcsin\left(2 \times \frac{\sqrt{1.86^2 - 2 \times 0.140^2}}{4}\right) \simeq 1.18 \text{ rad} \simeq 67.6°$$

Exercise 3.1.8

Equating the 4-momentum invariants in the LS for the initial state and in the CMS for the final state, we have at the threshold

$$E_p^2 + E_\gamma^2 + 2E_p E_\gamma - p_p^2 - p_\gamma^2 - 2\boldsymbol{p}_p \cdot \boldsymbol{p}_\gamma = (M_p + M_\pi)^2$$

In the UHE regime we assume $E_p \approx p_p$ and then we get

$$2E_p E_\gamma (1 - \cos\theta) = (M_p + M_\pi)^2 - M_p^2$$

The threshold energy as a function of the scattering angle is then

$$E_{\text{th}}(\theta) = \frac{(M_p + M_\pi)^2 - M_p^2}{2E_\gamma(1 - \cos\theta)}.$$

The minimum value is obtained in the case of *head-on* scattering, $\theta = \pi$

$$E_{\text{th}}^{\min} = \frac{(M_p + M_\pi)^2 - M_p^2}{4E_\gamma} \simeq 6.8 \times 10^{19} \text{ eV}.$$

Exercise 3.1.9

Using the relativistic invariants we have

$$(M_p + M_n)^2 = (E_d + E_\gamma)^2 - (\boldsymbol{p}_d + \boldsymbol{p}_\gamma)^2 = M_d^2 + 2E_\gamma E_d - 2\boldsymbol{p}_d \cdot \boldsymbol{p}_\gamma.$$

From the momentum conservation we can write $\boldsymbol{p}_d = -\boldsymbol{p}_\gamma$ and then

$$(M_p + M_n)^2 = M_d^2 + 2E_\gamma(E_d - p_d) \simeq M_d^2 + 2E_\gamma M_d \simeq (M_d + E_\gamma)^2,$$

where, in the last two steps, we have considered that both the deuteron recoil momentum the photon energy (both $O(\text{MeV})$) are negligible with respect to the deuteron mass. Hence we can write

$$M_d \simeq M_p + M_n - E_\gamma.$$

Assuming that the proton and neutron masses have negligible errors, it follows that $\Delta M_d = \Delta E_\gamma$, and finally we get

$$M_d = 1875.607 \pm 0.005 \text{ MeV}/c^2$$

Exercise 3.1.10

1. The total CMS energy is

$$E^* = \sqrt{2M_p^2 + 2M_p E_{\bar{p}}} \simeq 2.08 \text{ GeV}$$

were we used $E_{\bar{p}} = \sqrt{p_{\bar{p}}^2 + M_p^2} = 1.37$ GeV. The kaons in final state are produced back-to-back at $90°$ in the CMS. Their energies are $E_K^* = E^*/2$ and the momenta are

$$p_K^* = \sqrt{(E^*/2)^2 - M_K^2} \simeq 0.92 \text{ GeV}/c.$$

Therefore in the CMS $p_T^* = p_K^*$ and $p_L^* = 0$. Using the Lorentz transformation to the LS we get

$$(p_K)_T = p_T^* \simeq 0.92 \text{ GeV}/c$$

$$(p_K)_L = \gamma_{CM}[p_L^* + \beta_{CM}E_K^*] = 0 + \frac{p_{\bar{p}}}{E^*}\sqrt{(p_K^*)^2 + M_K^2} \simeq 0.50 \text{ GeV}/c$$

The kaon energy in the LS is then

$$E_K = \sqrt{[(p_K)_T^2 + (p_K)_L^2] + M_K^2} \simeq 1.16 \text{ GeV}$$

and the production angle is

$$\theta_K = \arctan\left[\frac{(p_K)_T}{(p_K)_L}\right] \simeq 61.5°$$

The kaons we are detecting have $\beta\gamma = p_K/M_K = 2.1$ and we can then assume that their energy loss in the gas is $\left(\frac{dE}{dx}\right)_{ion} = 2 \frac{\text{MeV}}{\text{g cm}^{-2}}$. The number of electron-ion pairs in each detector turns out to be

$$n = \frac{1}{I}\left(\frac{dE}{dx}\right)_{ion} \rho\, d\, \epsilon_p\, \epsilon_c \simeq \frac{2\ 10^6}{15} \times 2\ 10^{-3} \times 10 \times 0.20 \times 0.30 \simeq 160.$$

Exercise 3.1.11

The mass of the particle is obtained from the invariant mass of the two muons. Their energies in the LS are

$$E_1 = \sqrt{p_1^2 + m_\mu^2} \simeq \sqrt{45^2 + 106^2} \simeq 115 \text{ MeV}$$

$$E_2 = \sqrt{p_2^2 + m_\mu^2} \simeq p_2 = 30 \text{ GeV}.$$

The square of the total 4-momentum is the invariant mass of the system. The total energy and momentum are

$$E_t = E_1 + E_2 \simeq 0.115 + 30 \simeq 30.12 \text{ GeV}$$

$$|\boldsymbol{p}_t| = |\boldsymbol{p}_1 + \boldsymbol{p}_2| = |\boldsymbol{p}_2| - |\boldsymbol{p}_1| = 30 - 0.045 = 29.96 \text{ GeV/c}$$

Hence for the mass we have

$$M = \sqrt{E_t^2 - p_t^2} = \sqrt{30.12^2 - 29.96^2} = 3.10 \text{ GeV/c}^2.$$

The particle is the J/ψ-meson.

Exercise 3.1.12

At the threshold we have

$$E_p^2 + E_\gamma^2 + 2E_p E_\gamma - p_p^2 - p_\gamma^2 - 2\boldsymbol{p}_p \cdot \boldsymbol{p}_\gamma = (M_p + 2m_e)^2,$$

and for $E_p \simeq p_p$:

$$2E_p E_\gamma (1 - \cos\theta) = (M_p + 2m_e)^2 - M_p^2 \simeq 4M_p m_e.$$

Substituting $E_{\gamma\text{CMB}}$ to E_γ, the threshold energy as a function of θ turns out to be

$$E_{\text{th}}(\theta) \simeq \frac{2M_p m_e}{E_{\gamma\text{CMB}}(1 - \cos\theta)}.$$

The minimum value is obtained for $\theta = \pi$ (*head on* scattering) and is

$$E_{\text{th}}^{\text{min}} = \frac{M_p m_e}{E_{\gamma\text{CMB}}} \simeq 0.5 \times 10^{18} \text{ eV}.$$

Exercise 3.1.13

(a) Denoting by s the square of the total energy at LHC and with E_{Lab} the energy in fixed target pp interactions, we require

$$s = 2\,m\,E_{\text{Lab}}$$

where m is the proton mass. Here we have assumed that protons are ultra-relativistic. Hence we get

$$E_{\text{Lab}} = \frac{s}{2\,m} \simeq \frac{(13\ 10^{12})^2\ \text{eV}^2}{2 \times 0.94\ 10^9\ \text{eV}} \simeq 9 \times 10^{16}\ \text{eV}$$

(b) Denoting by v the insect velocity we have

$$\frac{1}{2} M v^2 = E_{\text{Lab}},$$

since in the ultra-relativistic limit the proton kinetic energy is almost equal to its total energy. Then we get

$$v = \sqrt{\frac{2\,E_{\text{Lab}}}{M}} \simeq \sqrt{\frac{2 \times 9\ 10^{16} \times 1.6\ 10^{-19}\ \text{J}}{0.25\ 10^{-3}\ \text{kg}}} \simeq 11\ \text{m/s} \simeq 39\ \text{km/h}$$

Exercise 3.1.14

(a) Consider the Lorentz transformation between the reference systems **K'** and **K**. Denoting by β the velocity of **K'** with respect to **K**, we have

$$E = \gamma(E' + \beta p'_\parallel), \quad p_\parallel = \gamma(p'_\parallel + \beta E').$$

Hence we get

$$E \pm p_\parallel = \gamma(1 \pm \beta)(E' \pm p'_\parallel),$$

$$\frac{E + p_\parallel}{E - p_\parallel} = \frac{1 + \beta}{1 - \beta} \times \frac{E' + p'_\parallel}{E' - p'_\parallel}$$

Using the definition of rapidity we finally obtain

$$y = y' + \ln\sqrt{\frac{1 + \beta}{1 - \beta}}$$

(b) The maximum (minimum) rapidity is obtained for the elastic scattering, $pp \to pp$, at $\theta = 0°$ ($\theta = 180°$). The maximum value is then for $p_\parallel = p$ and p equal to the beam momentum ($\simeq 6.5$ TeV)

$$y_{\text{max}} = \frac{1}{2}\ln\frac{E + p}{E - p} = \frac{1}{2}\ln\frac{(E + p)^2}{E^2 - p^2} = \ln\frac{E + p}{m} = \ln\frac{6500 + 6500}{0.94} \simeq 9.5$$

and $y_{\text{min}} = -y_{\text{max}}$.

(c) If θ is the scattering angle, we have $p_\parallel = p \cos\theta$. In the ultra-relativistic limit $E \simeq p$ and we get

$$y \simeq \frac{1}{2} \ln \frac{E(1 + \cos\theta)}{E(1 - \cos\theta)} = \frac{1}{2} \ln \frac{\cos^2\theta/2}{\sin^2\theta/2} = -\ln\tan\frac{\theta}{2} = \eta$$

(d) At $90°$ rapidity and pseudorapidity are identical: $y = \eta = 0$. At $1°$ we have

$$y = \frac{1}{2} \ln \frac{\sqrt{p^2 + m^2} + p \, \cos 1°}{\sqrt{p^2 + m^2} - p \, \cos 1°} = \frac{1}{2} \ln \frac{\sqrt{6500^2 + 0.94^2} + 6500 \, \cos 1°}{\sqrt{6500^2 + 0.94^2} - 6500 \, \cos 1°} \simeq 4.74123$$

$$\eta = -\ln\tan 0.5° \simeq 4.74134.$$

Therefore the difference is of the order of 1 over 10^5.

Exercise 3.1.15

For a particle moving along the x direction and emitting a decay particle at angle θ after a (proper) time t, we have

$$\Delta = ct\beta\gamma \sin\theta = ct\frac{p}{m} \sin\theta$$

Denoting CMS quantities with $*$ and with no index the ones in the LS, we obtain from the Lorentz transformations

$$p_y = p_y^*, \quad \epsilon = \gamma(\epsilon^* + \beta p_x^*)$$

$$p \sin\theta = p^* \sin\theta^*, \quad \epsilon = \gamma\epsilon^*(1 + \beta\beta^* \cos\theta^*),$$

where β^* is the velocity of the emitted particle in the CMS ($= p^*/\epsilon^*$). From their ratio we get

$$\beta\gamma \sin\theta = \frac{\beta^* \sin\theta^*}{1 + \beta\beta^* \cos\theta^*}$$

In the ultra-relativistic limit for both particles ($\beta \to 1$, $\beta^* \to 1$) we have

$$\beta\gamma \sin\theta \to \frac{\sin\theta^*}{1 + \cos\theta^*} = \tan\frac{\theta^*}{2}$$

and finally obtain

$$\Delta \to ct \tan\frac{\theta^*}{2}$$

which proves that the impact parameter is independent from the particle momentum.
    The mean value of the impact parameter for $t = \tau$ is

$$\langle \Delta \rangle = c\tau \int_0^{\pi} \tan \frac{\theta^*}{2} \frac{\sin \theta^* d\theta^*}{2} = c\tau \frac{(x - \sin x)|_0^{\pi}}{2} = \frac{\pi}{2} c\tau$$

For the D^+ decay we get

$$\langle \Delta \rangle \simeq 1.57 \times 3 \ 10^8 \ \text{m/s} \times 1.04 \ 10^{-12} \ \text{s} \simeq 490 \ \mu\text{m}$$

For a CMS angle $\theta^* = 90°$ we obtain

$$\Delta \simeq 3 \ 10^8 \ \text{m/s} \times 1.04 \ 10^{-12} \ \text{s} \times \tan 45° \simeq 312 \ \mu\text{m}$$

Exercise 3.1.16

The mass of the parent particle is the invariant mass of the two muons. We have

$$M^2 = P^2 = (p_1 + p_2)^2 = m_1^2 + m_2^2 + 2E_1E_2 - 2|\boldsymbol{p_1}||\boldsymbol{p_2}| \cos \theta,$$

where p_i and $\boldsymbol{p_i}$ are respectively the 4-momentum and the momentum of particle i (=1, 2). In the ultra-relativistic limit, holding for both muons, we get

$$M^2 = m_1^2 + m_2^2 + 4E_1E_2 \sin^2 \frac{\theta}{2} . \tag{3.4}$$

Substituting $m_1 = m_2 = m_\mu$ and considering that the muon mass is negligible with respect to the energies of both muons, we get

$$M = \sqrt{2m_\mu^2 + 4E_1E_2 \ \sin^2 \frac{\theta}{2}} \simeq 2\sqrt{E_1E_2} \ \sin \frac{\theta}{2} \simeq$$

$$\simeq 2\sqrt{7.4 \times 2.6} \ \sin \frac{42°}{2} \simeq 3.1 \ \text{GeV}/c^2$$

The mass value corresponds to the one of J/ψ.

The momentum of the particle can be obtained from the Carnot theorem

$$p = \sqrt{p_1^2 + p_2^2 + 2p_1p_2 \cos \theta} = \sqrt{7.4^2 + 2.6^2 + 2 \times 7.4 \times 2.6 \ \cos 42°} \simeq 9.5 \ \text{GeV}/c$$

and its energy is then $E = \sqrt{p^2 + M^2} \simeq 10 \ \text{GeV}$.

Substituting $E_2 = E - E_1$ in (3.4), we can express the opening angle as a function of E and E_1:

$$\sin \frac{\theta}{2} = \sqrt{\frac{M^2 - m_1^2 - m_2^2}{4E_1(E - E_1)}}$$

This expression is minimum for $E_1 = E/2$

$$\left(\sin\frac{\theta}{2}\right)_{\min} = \frac{\sqrt{M^2 - m_1^2 - m_2^2}}{E} \simeq \frac{M}{E} \simeq \frac{3.1}{10} = 0.31$$

which corresponds to an opening angle of 36°. The energy of the muons is then 10/2 = 5 GeV.

Exercise 3.1.17

(a) The CMS energy of a particle emitted in a two-bosy decay is ($M = m_\pi$)

$$\epsilon_\nu^* = \frac{m_\pi^2 + m_\nu^2 - m_\mu^2}{2m_\pi} = \frac{m_\pi^2 - m_\mu^2}{2m_\pi} \simeq \frac{0.140^2 - 0.106^2}{2 \times 0.140} \simeq 30 \text{ MeV}$$

(b) For a 200 GeV pion the Lorentz factor and the velocity are $\gamma = p/m_\pi \simeq 200/0.140 \simeq 1429$ and $\beta \approx 1$ respectively. Transforming the neutrino energy to the LS we get

$$E_\nu = \gamma(\epsilon_\nu^* + \beta p_\nu^* \cos\theta^*) = \gamma\epsilon_\nu^*(1 + \beta\cos\theta^*)$$

where θ^* is the neutrino emission angle in the rest frame. The maximum energy is obtained for $\theta^* = 0$ and is

$$E_\nu(\max) = \gamma\epsilon_\nu^*(1 + \beta) \simeq 2\gamma\epsilon_\nu^* \simeq 2 \times 1429 \times 30 \text{ MeV} \simeq 85.7 \text{ GeV}$$

(c) Consider neutrinos emitted at $\theta^* = 90°$ in the CMS. Their energy in the LS is

$$E_\nu(\theta^* = 90°) = \gamma\epsilon_\nu^*(1 + \beta\cos 90°) = \gamma\epsilon_\nu^* = \frac{E_\nu(\max)}{2} \simeq 42.9 \text{ GeV}.$$

Therefore forward emitted neutrinos have energies larger than this value.
(d) Using the relationship between angles under a Lorentz transformation we have

$$\tan\theta = \frac{\sin\theta^*}{\gamma(\cos\theta^* + \beta)}$$

For neutrinos in the forward hemisphere in the CMS the maximum angle corresponds to $\theta^* = 90°$

$$\tan\theta_{\max} = \frac{1}{\gamma\beta} \simeq \frac{1}{\gamma} \simeq 0.00070$$

which is an angle of 0.04°.

Exercise 3.1.18

If p and m are respectively the momentum and mass of the neutron, the distance it has to cover in a mean lifetime is

$$L = \gamma\beta c\tau = \frac{p}{m}\frac{cT_{1/2}}{\ln 2},$$

Fig. 3.1 Feynman diagram
for $\gamma + \gamma \rightarrow e^+ + e^-$

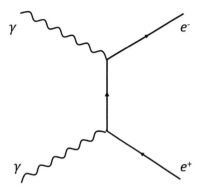

where $L (\simeq 5000 \times 365 \times 24 \times 60\, c \times$ minutes$)$ is the distance of the source. Hence we get

$$p \simeq \frac{L \times \ln 2}{T_{1/2}} \times m \simeq \frac{2.6\ 10^9\ \text{min} \times 0.693}{10\ \text{min}} \times m \simeq 1.8\ 10^8 \times 0.940\ \text{GeV} \simeq 1.7\ 10^{17}\ \text{eV}/c$$

Neutrons with such momentum are ultra-relativistic and thus also their energy has the same value.

Exercise 3.1.19

(a) The lowest order Feynman diagram is shown in Fig. 3.1. The amplitude is proportional to α and the cross section to α^2.

(b) Let us write $(E, \boldsymbol{K})$ the 4-momentum of the photon from the source and $(\epsilon, \boldsymbol{k})$ the one of the *CMB* photon, in the LAB system. $\boldsymbol{K}$ and $\boldsymbol{k}$ are opposite (head-on) and, considering also that they are massless, their sum is $E - \epsilon$. At the threshold we have:

$$P^2 = (E + \epsilon)^2 - (\boldsymbol{K} + \boldsymbol{k})^2 = (E + \epsilon)^2 - (E - \epsilon)^2 = (2m_e)^2$$

$$4E\epsilon = 4m_e^2,$$

and therefore the minimum photon energy is

$$E = \frac{m_e^2}{\epsilon} \simeq \frac{(0.511 \times 10^6)^2}{10^{-3}} \simeq 2.6 \times 10^{14}\ \text{eV}$$

(c) Denoting by M the invariant mass, at the threshold the Lorentz factor of the CMS system is

$$\gamma = \frac{E + \epsilon}{M} = \frac{E + \epsilon}{2m_e} \simeq 2.5 \times 10^8$$

3.2 Interaction of Radiation with Matter

Exercise 3.2.1

The number of photons as a function of the matter thickness x is

$$N(x) = N_0 \, e^{-\mu x}$$

where $\mu = 0.04 \text{ cm}^2/\text{g}$ for lead and x is expressed in g/cm^2. To halve the number of photons we require

$$\frac{N_0}{2} = N_0 e^{-\mu x_{1/2}} \implies x_{1/2} = \frac{\ln 2}{\mu} \simeq 17.33 \text{ g/cm}^2.$$

Using $\rho = 11.3 \text{ g/cm}^3$, we have $l_{1/2} = \frac{x_{1/2}}{\rho} = 1.53$ cm.
   For a 5% photon survival we have

$$0.05 \, N_0 = N_0 \, e^{-\mu x} \implies x = -\frac{1}{\mu} \ln(0.05) \simeq 75 \text{ g/cm}^2,$$

and then $l_{5\%} = 6.63$ cm.

Exercise 3.2.2

In the high energy limit ($E_\gamma \gg m_e$) the absorption coefficient for pair production is

$$\mu = \left(\frac{7}{9}\right) X_0^{-1} \simeq 1.4 \text{ cm}^{-1}$$

The cross section can be obtained from the absorption coefficient μ using the relationship

$$\mu = n\sigma, \quad \text{where} \quad n = \frac{\rho N_A}{A}.$$

Using $A = 207, \rho = 11.3 \text{ g/cm}^3$ and the Avogadro number $N_A = 6.02 \cdot 10^{23} \text{ mole}^{-1}$ we get

$$\sigma = \frac{A}{N_A} \frac{\mu}{\rho} \simeq 4.2 \cdot 10^{-23} \text{ cm}^2 = 42 \text{ b}$$

Exercise 3.2.3

Neglecting the momentum loss in the slab (see below), the radius of curvature of the muon is

$$R = \frac{p}{0.3 \, B} \simeq 33.3 \text{ m},$$

where, in this equation, B, R and p are given in Tesla, meter and GeV/c respectively. The muon deflection angle θ is equal to the angle of the radius at the exit with respect to the slab. For small angles the circular segment can be approximated to the slab thickness (see figure) and then we can write

$$\theta \simeq \arcsin\left(\frac{L}{R}\right) \simeq 0.015 \text{ rad} \simeq 0.86°$$

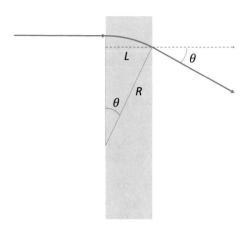

Considering its initial energy the muon energy loss rate corresponds to the one of a minimum ionising particle, $\left(-\frac{dE}{dx}\right) \simeq 1.4$ MeV cm^2/g. The energy loss is then

$$\Delta E = \left(-\frac{dE}{dx}\right) \times \rho l \simeq 550 \text{ MeV}.$$

where $l = R\theta \simeq L \simeq 50$ cm. Since $\frac{\Delta E}{E} = \frac{\Delta p}{p}$, it follows that the muon momentum after the slab is

$$p' = p - \Delta p \simeq p - \Delta E \simeq 19.5 \text{ GeV}/c.$$

The multiple scattering dispersion in the plane of the figure is given by

$$\sqrt{\langle \theta_s^2 \rangle} = \frac{E_s}{\sqrt{2}\bar{p}\beta}\sqrt{\frac{l}{X_0}} \simeq 4 \text{ mrad} \simeq 0.23°$$

where $E_s \simeq 20$ MeV is the multiple scattering constant and $\bar{p} = \sqrt{pp'} \simeq 19.75$ GeV/c (see Exercise 3.2.8). The factor $\sqrt{2}$ at denominator converts the spatial dispersion angle to the plane angle ($\theta_{proj}^2 = \theta_{space}^2/2$).

Exercise 3.2.4

The Compton scattering cross section in the low energy limit, $E_\gamma \ll m_e c^2$, is given by the Thomson cross section

$$\sigma = \frac{8}{3}\pi r_0^2,$$

where r_0 is *classical electron radius* $[= e^2/(4\pi\epsilon_0 mc^2) \simeq 2.8$ fm]. Hence we have

$$\lambda = \frac{1}{\sigma n} = \frac{1}{\frac{8}{3}\pi r_0^2 N_A \frac{Z}{A} \rho} \simeq 4.52 \text{ cm}$$

Exercise 3.2.5

Using the equation $p[\text{GeV}/c] = 0.3 \times B[\text{T}] \times R[\text{m}]$ and writing the sagitta as $s \simeq L^2/(8R)$, valid for $R \gg s$, the electron momentum is

$$p = 0.3 \, B \, \frac{L^2}{8s} \simeq 0.3 \times 0.1 \, \frac{0.03^2}{8 \times 0.002} \simeq 1.7 \text{ MeV}/c.$$

The kinetic energy of the electron is $T = \sqrt{p^2 + m^2} - m \simeq 1.3$ MeV.

The 4-momentum conservation in the Compton scattering can be written as

$$E_\gamma + m = E'_\gamma + E \qquad \boldsymbol{k} = \boldsymbol{k'} + \boldsymbol{p},$$

being $(E_\gamma, \boldsymbol{k})$ and $(m, \boldsymbol{0})$ the initial 4-momenta of the photon and electron, and $(E'_\gamma, \boldsymbol{k'})$ and $(E, \boldsymbol{p})$ the final ones. Squaring $\boldsymbol{k} - \boldsymbol{p} = \boldsymbol{k'}$ and solving in E_γ, we get for the initial photon energy[6]

$$E_\gamma = \frac{p^2 - T^2}{2(p \, cos\phi - T)} \simeq 1.6 \text{ MeV}.$$

The scattered photon energy is

$$E'_\gamma = E_\gamma - T \simeq 0.3 \text{ MeV}$$

Exercise 3.2.6

The mean number of pairs created by a single pion is

$$n = \frac{\left(-\frac{dE}{dx}\right)_{\text{ion}} \rho \, d}{\langle I \rangle}.$$

In this exercise, as in many others in this book, the value of the ionization loss rate is not given for the specific case (particle, material, etc.). Most of the cases refer to relativistic singly charged particles. To help making a correct estimate one should have in mind the main features of $(-dE/dx)_{\text{ion}}$ that can be easily deduced from a figure of this function, e.g., as reported in the PDG Review of Particle Physics [1]. These features can be summarized as follows:

• the minimum of $(-dE/dx)_{\text{ion}}$ is at $\beta\gamma \approx 3$. The differences in the minimum ionization loss rate among the different materials is modest, because it is mainly determined by the ratio Z/A: they change from ≈ 1.2 MeV g^{-1} cm^2 for Pb up to

---

[6]It should not surprise that the scattered electron has a momentum larger than the initial photon energy. In fact the photon is scattered backward at an angle of about 108°.

≈ 2 MeV g^{-1} cm^2 for He. The only exception is hydrogen, whose Z/A is 2 and about twice w.r.t all the other elements, which has a minimum ionization energy loss rate of ≈ 4 MeV g^{-1} cm^2.

• In the relativistic and ultra-relativistic regimes, the increase of $(-dE/dx)_{ion}$ with $\beta\gamma$ is very small and in some case negligible. To have a reference number, there is a factor of about 1.5 with respect to the minimum ionization in the $\beta\gamma$ range from 3 to 10,000.

• The previous consideration is actually true only for solid and liquid materials. In these materials the energy loss is modified at increasing $\beta\gamma$ by the so-called "density effect". Instead for gases this effect is negligible and the increase of $(-dE/dx)_{ion}$ with $\beta\gamma$ is somewhat larger (in the $\beta\gamma$ range from 3 to 10,000 a factor about 2).

20 GeV pions have $\beta\gamma \simeq 140$ and the medium in the counter is a gas (whose composition is not given). Taking into account the fact that the typical gases used in ionization counters have a minimum ionization of $1.5 \div 2$ MeV g^{-1} cm^2 and a contribution due to the relativistic increase in gases, we can assume an energy loss rate

$$\left(-\frac{dE}{dx}\right)_{ion} \simeq 2 \frac{\text{MeV cm}^2}{\text{g}}.$$

Using $d = 1$ cm, $\rho = 1.8 \times 10^{-3}$ g/cm^3 and $\langle I \rangle = 15$ eV, we get $n = 240$ pairs. Hence we have for the current

$$I_{out} = I_0 \times n = 2.4 \text{ mA}$$

Exercise 3.2.7

The velocity for protons and pions are

$$\beta_p = \frac{p}{\sqrt{p^2 + M_p^2}} \simeq 0.983 \qquad \beta_\pi = \frac{p}{\sqrt{p^2 + M_\pi^2}} \simeq 0.9996$$

Hence they are related as

$$\beta_\pi > \beta_p > \frac{1}{n_1} = 0.952$$

So the first Cherenkov detector is sensitive to both particle types. To get the beam separation we then require a refractive index in the second detector allowing the detection of the faster particle only

$$\beta_\pi > \frac{1}{n_2} > \beta_p$$

from which we get

$$1.0004 < n_2 < 1.017.$$

Exercise 3.2.8

500 MeV/c muons have $\beta\gamma \simeq 4.7$ and so they are close to the ionization minimum. For the copper slab we can assume

$$\left(-\frac{dE}{dx}\right)_{\text{ion}} = 1.4 \frac{\text{MeV}}{\text{g cm}^{-2}}.$$

The thickness to stop the muon beam is the range for these muons. A simple estimate can be done assuming that the energy loss is constant along the particle trajectory

$$R = \frac{1}{\rho} \int_0^T \frac{dT}{(-dE/dx)_{\text{ion}}} \simeq$$

$$\simeq \frac{1}{\rho} \frac{T}{(-dE/dx)_{\text{ion}}} = \frac{\sqrt{p^2 + M^2} - M}{\rho\,(-dE/dx)_{\text{ion}}} \simeq \frac{405\ \text{MeV}}{9 \times 1.4\ \text{MeV/cm}} \simeq 32\ \text{cm},$$

where T and p are the initial muon kinetic energy and momentum respectively, and M is their mass. A better value for the range can be obtained from the graph R/M versus $\beta\gamma$ shown in the figure below, taken from the PDG Review of Particle Physics [1]. From this figure we deduce for an element (Fe) close to the copper $R/M \simeq 2300\ \text{g cm}^{-2}/\text{GeV}$. Substituting to M the muon mass we get $R \simeq 27\ \text{cm}$.

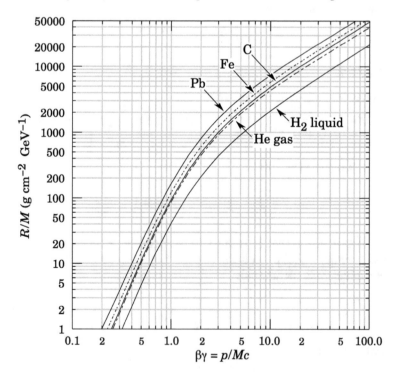

The kinetic energy lost in a 10 cm slab is

$$\Delta T = \left(-\frac{dE}{dx} \right)_{\text{ion}} \rho \, d = 126 \text{ MeV}.$$

Hence the mean energy of the muons after the slab crossing is

$$T' = T - \Delta T \simeq 279 \text{ MeV}.$$

The multiple scattering angle at the exit has to be calculated taking into account the ionization energy loss in the slab, because this loss is not negligible $\Delta T / T \simeq 126/405 \simeq 31\%$. The calculation has to be done as follows

$$d\theta^2 = \left(\frac{E_s}{p\beta} \right)^2 \frac{dx}{X_0} = \left(\frac{E_s}{p\beta} \right)^2 \frac{dp\beta}{X_0 \, (-dp\beta/dx)_{\text{ion}}}.$$

where X_0, converted to g/cm^2, is $1.4 \times 9 = 12.6$ g/cm^2 and $(-dp\beta/dx)_{\text{ion}}$, the $p\beta$ loss rate, can be obtained from the ionization energy loss rate as

$$\left(-\frac{dp\beta}{dx} \right)_{\text{ion}} = \left(-\frac{dE}{dx} \right)_{\text{ion}} \frac{dp\beta}{dT}.$$

We can write

$$p\beta = \frac{p^2}{E} = \frac{(T+M)^2 - M^2}{T+M} = T \left(1 + \frac{M}{T+M} \right) \tag{3.5}$$

hence we get

$$\left(-\frac{dp\beta}{dx} \right)_{\text{ion}} = \left(-\frac{dE}{dx} \right)_{\text{ion}} \left[1 + \left(\frac{M}{T+M} \right)^2 \right] = \left(-\frac{dE}{dx} \right)_{\text{ion}} [1 + \epsilon(T)]$$

The function $\epsilon(T)$ is ~4% for the entrance energy and ~7.5% for the exit energy. For an estimate of the scattering angle (within an accuracy of less than 10%) we can neglect such function in the previous expression and calculate the r.m.s. scattering angle as

$$\theta_s^2 = \langle \theta^2 \rangle = \int_{(p\beta)_i}^{(p\beta)_f} d\theta^2 = \frac{E_s^2}{X_0 \, (-dE/dx)_{\text{ion}}} \int_{(p\beta)_i}^{(p\beta)_f} \frac{dp\beta}{(p\beta)^2} =$$

$$= \frac{E_s^2}{X_0 \, (-dE/dx)_{\text{ion}}} \frac{(p\beta)_f - (p\beta)_i}{(p\beta)_f (p\beta)_i}, \tag{3.6}$$

where $(p\beta)_i \simeq 489$ MeV/c and $(p\beta)_f \simeq 356$ MeV/c are the $p\beta$ value corresponding to the entrance and exit of the muons. In the previous expression we have assumed constant the energy loss rate within the integration range. Hence we obtain

$$\theta_s^2 = \frac{20^2}{12.6 \cdot 1.4} \frac{489 - 356}{489 \cdot 356} \simeq 0.017$$

and

$$\theta_s \simeq 0.132 \text{ rad} \simeq 7.5°.$$

Equation (3.6) allows to get a simple rule to calculate the multiple scattering angle to be used in case of sizeable energy loss. In fact, considering that we have $(p\beta)_f - (p\beta)_i \approx (-dE/dx)_{ion} \times d$, the r.m.s. scattering angle can be written as in the case of $p\beta$ constant

$$\theta_s = \left(\frac{E_s}{[p\beta]}\right)\sqrt{\frac{d}{X_0}}$$

replacing $p\beta$ with the geometric mean $[p\beta] = \sqrt{(p\beta)_f(p\beta)_i} \simeq 417$ MeV/c.

Exercise 3.2.9

The energy of the photons which are incident on the silver foil is

$$E_\gamma = \frac{hc}{\lambda} \simeq \frac{6.28 \times 197 \; 10^6 \times 10^{-6} \text{ eV nm}}{200 \text{ nm}} \simeq 6.2 \text{ eV}.$$

To have the photoelectric effect the photon energy must fulfill the condition $E_\gamma > W$. With $W = 4.73$ eV the photoelectric process is allowed. The electron kinetic energy is $E_K = E_\gamma - W = 1.47$ eV.

Exercise 3.2.10

According to the Heitler toy model, the depth T at which the shower reaches the maximum development is given by the equation $2^T = E_0/E_{crit}$. Hence we have

$$T = \frac{\log_{10}(E_0/E_{crit})}{\log_{10} 2} = \frac{\log_{10}(100 \text{ GeV}/80 \text{ MeV})}{\log_{10} 2} \simeq 10.3$$

where T is given in units of radiation lengths. Hence the air thickness is

$$X_{max} = T \times X_0 = 10.3 \times 37 \text{ g/cm}^2 \simeq 380 \text{ g/cm}^2$$

Exercise 3.2.11

The photons emitted in the e^+e^- annihilation at rest have energy $E_\gamma = M/2 = m_e$, being $M = 2m_e$ and m_e the electron mass. In the Compton process the scattered photons have energy

$$E'_\gamma = \frac{E_\gamma}{1 + E_\gamma/m_e(1 - \cos\theta)}$$

from which it follows that the extremal electron kinetic energies are for $\cos\theta = 1$ and $\cos\theta = -1$

$$(T_e)_{\min} = E_\gamma - (E'_\gamma)_{\max} = E_\gamma - E_\gamma = 0$$

$$(T_e)_{\max} = E_\gamma - (E'_\gamma)_{\min} = E_\gamma - \frac{E_\gamma m_e}{2E_\gamma + m_e} = \frac{2}{3}m_e \simeq 0.34 \text{ MeV}$$

Exercise 3.2.12

Assuming an energy loss rate of $2 \text{ MeV}/(\text{g} \cdot \text{cm}^{-2})$, the minimum kinetic energy for a vertical muon to reach ground[7] is

$$T_{\min} \simeq \left(-\frac{dE}{dx}\right)_{\text{ion}} \times \Delta x \simeq 2 \frac{\text{MeV}}{\text{g} \cdot \text{cm}^{-2}} \times 1030 \text{ g} \cdot \text{cm}^{-2} \simeq 2.1 \text{ GeV}$$

The number of ionized electrons are

$$N_e = \frac{T_{\min}}{\langle I \rangle} \simeq \frac{2.1 \times 10^9 \text{ eV}}{10 \text{ eV}} \simeq 10^8.$$

Exercise 3.2.13

We notice first that the plate thickness is much smaller than a radiation length so that we can neglect the electron energy loss. In this condition the r.m.s. scattering angle is simply

$$\theta_s = \sqrt{\langle \theta^2 \rangle} \simeq \frac{E_s}{E_0}\sqrt{\frac{x}{X_0}}$$

where E_0 is the electron energy, $E_s \simeq 20 \text{ MeV}$ and $x = X_0/20$. Considering the electron bremsstrahlung the dispersion angle is about

$$\theta_b \simeq \frac{m_e}{E_0}$$

where m_e is the electron mass. We have

$$\theta_s = \frac{20}{1000}\sqrt{\frac{1}{20}} \simeq 4.5 \text{ mrad} \gg \theta_b = \frac{0.5}{1000} \simeq 0.5 \text{ mrad}$$

and then the angular distribution is dominated by the multiple scattering.

---

[7]The muon mean decay pathlength is $c\tau = 660$ m. This value does not contrast with the total pathlength $(10 \div 15 \text{ km})$ from the muon production to the sea level. At the minimum energy the mean decay pathlength in the Earth reference system is $L_\mu = \beta\gamma c\tau \simeq E_{\min}/m_\mu \times c\tau \simeq 2100/106 \times 660 \text{ m} \simeq 13 \text{ km}$. This means that most of the muons above the minimum energy reach the ground.

Exercise 3.2.14

For muons ($E = 3$ GeV) in copper we can assume (see Exercise 3.2.6) a ionization energy loss rate

$$\left(-\frac{dE}{dx}\right)_{ion} = 1.8 \, \frac{\text{MeV}}{\text{g cm}^{-2}}.$$

In a $d = 10$ cm thick slab muons lose an energy

$$\Delta E = \left(-\frac{dE}{dx}\right)_{ion} \rho d = 0.16 \, \text{GeV},$$

and so we have $E \gg \Delta E$. The lateral beam broadening can be calculated assuming that the muon energy is unaffected by the slab crossing and we can write

$$\langle(\rho r)^2\rangle = \int_0^{x_S} x^2 d\theta^2 = \int_0^{x_S} dx \, \frac{x^2}{X_0} \left(\frac{E_s}{p\beta}\right)^2 \simeq \frac{x_S^3}{3X_0}\left(\frac{E_s}{p\beta}\right)^2$$

where $x_S = d\rho \simeq 90$ g/cm^2 is the slab mass thickness and $E_s = 20$ MeV is the scattering constant. Hence we have

$$\sqrt{\langle(\rho r)^2\rangle} \simeq x_S\sqrt{\frac{x_S}{3X_0}}\left(\frac{E_s}{p\beta}\right)$$

Since $E \gg M_\mu$, we can write $p\beta \approx E$ and finally get for the beam broadening

$$\sqrt{\langle r^2\rangle} \simeq 10\sqrt{\frac{90}{3 \times 13.3}}\left(\frac{20}{3000}\right) \simeq 0.1 \, \text{cm}.$$

Exercise 3.2.15

The quantity $z^2 R/M$ (where z is the charge in e units, R the range and M the mass of the particle) is a universal function of $\beta\gamma = p/M$. As an example an α-particle having a kinetic energy T_α has the same range of a proton with kinetic energy $T_p = T_\alpha/4$ (same $\beta\gamma$), because $z_\alpha^2/M_\alpha = 1/M_p$.

Exercise 3.2.16

Electrons having $E = 1$ GeV loose energy by bremstrahlung as

$$\left(-\frac{dE}{dx}\right)_{brem} = \frac{E}{X_0}.$$

Therefore the mean electron energy after crossing a plate of thickness x is

$$E(x) = E_0 \, \exp\left(-\frac{x}{X_0}\right). \tag{3.7}$$

X_0 is the aluminium radiation length whose inverse is

$$\frac{1}{X_0} \approx D\frac{Z^2}{A} \ln(183\ Z^{-1/3}) \simeq 3.8 \times 10^{-2}\ \text{cm}^2/\text{g}.$$

The radiated energy corresponds to the value of $E(x)$ in Eq. (3.7) at $x = \rho d \simeq 13.5$ g/cm^2, and then is

$$\langle E_\gamma \rangle = \Delta E = E_0 \left[1 - \exp\left(-\frac{x}{X_0}\right)\right] \simeq 0.40 \times E_0 = 400\ \text{MeV}.$$

Exercise 3.2.17

(a) In vacuum muons travel along a circular orbit whose radius is

$$R[\text{m}] = \frac{p[\text{GeV}/\text{c}]}{0.3\ B[\text{T}]} \simeq 16.7\ \text{m}$$

(b) Muons have an initial energy

$$E_0 = \sqrt{p_0^2 + m^2} = 511\ \text{MeV}$$

and a $\beta\gamma$ equal to $p/m \simeq 4.7$. Hence we can assume for the energy loss rate in the gas a value close to that of a minimum ionizing particle

$$\left(\frac{dE}{dx}\right)_{\text{ion}} \simeq 2\ \frac{\text{MeV}}{\text{g/cm}^2}.$$

The energy lost after a complete round is approximately[8]

$$\Delta E \simeq \left(\frac{dE}{dx}\right)_{\text{ion}} \rho \times 2\pi R \simeq 42\ \text{MeV}$$

and the final energy is $E_1 = E_0 - \Delta E \simeq 469$ MeV. The corresponding muon momentum is $p_1 = \sqrt{E_1^2 - m^2} \simeq 457$ MeV/c and so the radius of curvature after one round is $R_1 \simeq 15.2$ m.

Exercise 3.2.18

The scattered photon energy as a function of the scattering angle θ is

$$E_\gamma' = \frac{E_\gamma}{1 + \epsilon\ (1 - \cos\theta)}$$

---

[8]For an estimate it is not necessary to consider the actual trajectory of the muon which slightly differs from a circle.

where $\epsilon = E_\gamma/m \simeq 1$, for 0.5 MeV photons. If E, T and m are the energy, kinetic energy and mass of the scattered electron, from the energy conservation we have

$$E_\gamma + m = E'_\gamma + E$$

and then

$$T = E - m = E_\gamma - E'_\gamma = E_\gamma \frac{\epsilon (1 - \cos \theta)}{1 + \epsilon (1 - \cos \theta)}$$

The maximum energy is obtained when the photon is scattered backward ($\theta = \pi$) and hence we obtain for the Compton edge energy

$$T_{\max} = E_\gamma \frac{2\epsilon}{1 + 2\epsilon} \simeq \frac{2}{3} E_\gamma \simeq 0.33 \text{ MeV}$$

Exercise 3.2.19

The energy of the Compton scattered photon as a function of θ is

$$E'_\gamma = \frac{E_\gamma}{1 + \epsilon(1 - \cos \theta)}$$

where $\epsilon = E_\gamma/m$. The electron kinetic energy is then

$$T = E - m = E_\gamma - E'_\gamma = E_\gamma \frac{\epsilon (1 - \cos \theta)}{1 + \epsilon (1 - \cos \theta)}$$

This energy is maximum for $\theta = \pi$ and this value corresponds to the co-called 'Compton edge'

$$T_{\max} = E_\gamma \frac{2\epsilon}{1 + 2\epsilon} = E_\gamma \frac{2E_\gamma}{m + 2E_\gamma}.$$

Solving the equation in E_γ we have

$$E_\gamma = \frac{T_{\max} + \sqrt{T_{\max}(T_{\max} + 2m)}}{2}. \tag{3.8}$$

The three $T_{\max}$ values shown in the figure are about 0.22, 0.62 and 0.80 MeV. Knowing that in γ-transitions, neglecting the nucleus recoil, $Q_\gamma = E_\gamma$, from (3.8) we get

$$Q_\gamma(1) \simeq 0.37 \text{ MeV}, \quad Q_\gamma(2) \simeq 0.81 \text{ MeV}, \quad Q_\gamma(3) \simeq 1.0 \text{ MeV}$$

Exercise 3.2.20

(a) The muon velocity is

$$\beta = \frac{p}{E} = \frac{p}{\sqrt{p^2 + m^2}} = \frac{10}{\sqrt{10^2 + 0.106^2}} \simeq 0.999944$$

Thus Cherenkov effect is done because we have

$$\beta \simeq 0.999944 > \frac{1}{n} = \frac{1}{1.00029} \simeq 0.99971$$

(b) The Cherenkov opening angle is given by

$$\cos \theta_C = \frac{1}{n\beta} \simeq \frac{1}{1.00029 \times 0.999944} \simeq 0.99977$$

corresponding to an angle of $1.2°$.

(c) The number of Cherenkov photons per unit length in the visible bandwidth is

$$N_{ph}/L \approx z^2 \frac{\alpha}{c} \Delta\omega \sin^2 \theta_C \approx z^2 \, 750 \sin^2 \theta_C \; \text{cm}^{-1},$$

where $\Delta\omega$ corresponds to the visible and near UV bandwidth, where Cherenkov radiation is possible ($\hbar\Delta\omega \approx 2$ eV). 10 GeV muons produced at 10 km reach the sea level since their mean decay length is ($\tau_\mu \simeq 2.2 \, \mu$s):

$$l_\mu = \beta\gamma c\tau_\mu = \frac{p_\mu}{m_\mu} c\tau_\mu \simeq \frac{10}{0.106} \, 3 \, 10^5 \, 2.2 \, 10^{-6} \; \text{km} \simeq 62 \; \text{km},$$

and then emit Cherenkov photons along their whole pathlengths. For muons hitting normally the Earth surface ($\theta_Z = 0$) we have

$$N_{ph} \approx 750(1 - \cos^2 \theta_C) \times L \simeq 750(1 - 0.99977^2) \; \text{cm}^{-1} \times 10^6 \text{cm} \simeq 3.4 \, 10^5.$$

For angles $\theta_Z > 0$, the number of photons scales as $\sec \theta_Z$.

Exercise 3.2.21

From the 4-momentum conservation in the Compton scattering we have

$$E_\gamma + m = E'_\gamma + E \qquad \mathbf{k} = \mathbf{k'} + \mathbf{p}$$

where $(E_\gamma, \mathbf{k})$, $(E'_\gamma, \mathbf{k'})$ and $(E, \mathbf{p})$ are the 4-momenta of the incident photon, scattered photon and scattered electron respectively. We need to calculate the relationship between the initial photon and the scattered electron as a function of the electron angle ϕ. To get this we write

$$\mathbf{k'}^2 = (\mathbf{k} - \mathbf{p})^2 = k^2 + p^2 - 2kp \cos \phi = E_\gamma^2 + p^2 - 2E_\gamma p \cos \phi$$

$$\mathbf{k'}^2 = E_\gamma'^2 = (E_\gamma + m - E)^2 = (E_\gamma - T)^2$$

where T is the electron kinetic energy. Hence we have

$$E_\gamma = \frac{p^2 - T^2}{2(p \, \cos\phi - T)}$$

(a) An electron having an angle ϕ within the fibre acceptance releases its whole kinetic energy, because it has enough pathlength to come at rest. Thus the measured energy release corresponds to the kinetic energy of the electron at $\phi = 30°$ and we have for the source energy

$$E_\gamma \simeq \frac{2.46^2 - 2^2}{2 \cdot (2.46 \cdot \cos 30° - 2)} \simeq 7.9 \text{ MeV},$$

where we have used $p = \sqrt{(T + m)^2 - m^2} \simeq 2.46$ MeV/c.
(b) The Klein–Nishina cross section is

$$\frac{d\sigma}{d\Omega} = \frac{r_0^2}{2} \left(\frac{E_\gamma'}{E_\gamma}\right)^2 \left[\frac{E_\gamma'}{E_\gamma} + \frac{E_\gamma}{E_\gamma'} - \sin^2\theta\right] \tag{3.9}$$

where θ is the photon scattering angle. To get this angle we equate the photon and electron transverse momenta

$$p \, \sin\phi = E_\gamma' \, \sin\theta \qquad \Longrightarrow \qquad \sin\theta = \frac{p}{E_\gamma'} \sin\phi$$

The photon energy E_γ', corresponding to the electron emitted at 30°, can be derived from $E_\gamma' = E_\gamma - T$. Hence we have $E_\gamma'/E_\gamma \simeq 0.75$ and $\sin\theta \simeq 0.21$. Substituting these values in (3.9) we get

$$\frac{d\sigma}{d\Omega} \simeq \frac{2.8 \text{ fm}^2}{2} \times 0.75^2 \times (0.75 + 1/0.75 - 0.21^2) \simeq 4.5 \ 10^{-26} \text{ cm}^2/\text{sr}.$$

The cross section for all the accepted electrons is this differential cross section multiplied by the acceptance solid angle

$$\Delta\Omega = 2\pi \int_0^{15°} d\cos\theta = 2\pi(1 - \cos 15°) \simeq 6.28 \times (1 - 0.966) \simeq 0.21 \text{ sr}$$

Hence we have

$$\sigma_{\text{acc}} \simeq \frac{d\sigma}{d\Omega} \Delta\Omega \simeq 4.5 \ 10^{-26} \times 0.21 \simeq 9.5 \ 10^{-27} \text{cm}^2.$$

(c) The Compton absorption coefficient for the accepted electrons is

$$\mu_{acc} = N\frac{Z}{A}\rho\,\sigma_{acc} \simeq 6\ 10^{23} \times 0.5 \times 1 \times 9.5\ 10^{-27} \simeq 0.0029\ \text{cm}^{-1}$$

where we have taken into account that Z electrons per each atom contribute to the scattering. The number of detected electrons per incident photon is then

$$\frac{N_e}{N_\gamma} \simeq \frac{d}{\sin 30°}\mu_{acc} \simeq \frac{0.2}{0.5} \times 0.0029 \simeq 0.001.$$

Exercise 3.2.22

By definition $E(x) = E_0\,\exp(-x/X_0)$, where X_0 is the radiation length. The mean energy loss is then $\Delta E = E_0 - E(X_0) = E_0(1 - 1/e) = 0.63$ GeV.

3.3 Detection Techniques and Experimental Methods

Exercise 3.3.1

(1) The inverse of β is given by

$$\frac{1}{\beta} = \frac{1}{\sqrt{1 - \frac{1}{\gamma^2}}}.$$

Since $E \gg m$, we have $\gamma \gg 1$, so that can use the relationship, valid for $x \to 0$

$$\frac{1}{\sqrt{1 - x^2}} \simeq 1 + \frac{1}{2}x^2.$$

The difference between the time-of-flights of two particles having velocities β_1 and β_2 is

$$\Delta T = \frac{L}{\beta_1 c} - \frac{L}{\beta_2 c} \simeq \frac{L}{c}\left(1 + \frac{1}{2\gamma_1^2} - 1 - \frac{1}{2\gamma_2^2}\right) = \frac{L}{2c}\left(\frac{1}{\gamma_1^2} - \frac{1}{\gamma_2^2}\right) =$$

$$= \frac{L}{2c}\left(\frac{m_1}{E_1^2} - \frac{m_2}{E_2^2}\right) \simeq \frac{L}{2c}\left(\frac{m_1^2}{p_1^2} - \frac{m_2^2}{p_2^2}\right) = \frac{L}{2c}\frac{m_1^2 - m_2^2}{p^2},$$

having set $p_1 = p_2 = p$ in the last step.

(2) The difference in time-of-flight between pions and kaons is

$$\Delta T = \frac{L}{2c}\frac{\Delta m^2}{p^2} = \frac{3\ \text{m}}{2\ c}\frac{0.493^2 - 0.139^2}{1} \simeq \frac{3\ \text{m} \times 0.224}{2 \times 3 \times 10^8 \text{m/s}} \simeq 1.12\ \text{ns}$$

Using the time resolution requirement, $\Delta T = 4\sigma_t$, we obtain for the time resolution needed for each counter

$$\sigma \simeq \frac{1.12 \text{ ns}}{4\sqrt{2}} \simeq 0.2 \text{ ns},$$

where we have used the relationship $\sigma_t^2 = \sigma^2(T_1 - T_2) = \sigma^2(T_1) + \sigma^2(T_2) = 2\sigma^2$.
(3) When S1 and S2 are segmented and a third scintillator S3 is inserted in the middle, the system can be used as a spectrometer.

(a) The space resolution of each scintillator is $\sigma_y = 5 \text{ cm}/\sqrt{12} \simeq 1.44 \text{ cm}$ (y is the direction orthogonal to the beam in the figure). The lateral spread due to the multiple scattering (see Exercise 3.2.14), in the same direction, is:

$$\sigma_y \simeq \frac{x}{\sqrt{3}} \frac{E_s}{\sqrt{2} \, p\beta} \sqrt{\frac{x}{X_0}} \simeq \frac{1 \text{ cm}}{\sqrt{3}} \frac{14 \text{ MeV}}{1000 \text{ MeV} \, \beta} \sqrt{\frac{1}{40}} \simeq \frac{0.013}{\beta} \text{ mm},$$

which turns out to be negligible for both particles ($\beta_\pi \simeq 0.99$, $\beta_K \simeq 0.90$) with respect to the resolution.

(b) The sagitta is

$$s = 0.3 \frac{BL^2}{8 \, p} = \frac{0.3 \times 1 \text{ T} \times 9 \text{ m}^2}{8 \times 1 \text{ GeV/c}} \simeq 0.337 \text{ m} = 33.7 \text{ cm}.$$

The sagitta is measured as $s \simeq y_3 - (y_1 + y_2)/2$ and the uncertainty on each y_i is σ_y. Hence we have for the sagitta uncertainty

$$\sigma_s \simeq \sqrt{\frac{3}{2}} \sigma_y \simeq 1.76 \text{ cm}.$$

The relative error on the momentum measurement is finally

$$\frac{\Delta p}{p} \simeq \frac{\Delta s}{s} = \frac{1.76 \text{ cm}}{33.7 \text{ cm}} \simeq 5\%$$

Exercise 3.3.2

The radius R of the orbit at $t = t_o$ is

$$R[\text{m}] = \frac{p[\text{GeV/c}]}{0.3 \, B[\text{T}]} = \frac{0.3}{0.3 \times 0.5} \text{ m} \simeq 2 \text{ m}.$$

A 300 MeV/c muon ($\beta\gamma \simeq 3$) is at the minimum of the ionization loss rate. The medium is not specified but it may be a gas, considering its density. Let assume to be air for which the minimum ionization loss is $\approx 1.8 \text{ MeV g}^{-1} \text{ cm}^2$. For the energy loss in iron we use instead $\approx 1.5 \text{ MeV g}^{-1} \text{ cm}^2$. Under this assumption we have

$$\Delta E \simeq 1.5 \, \frac{\text{MeV}}{\text{g cm}^{-2}} \times \rho_{\text{Fe}} \times 2d_{\text{Fe}} + 1.8 \, \frac{\text{MeV}}{\text{g cm}^{-2}} \times \rho_{\text{air}} \times 2\pi R \simeq$$

$$= 1.5 \times 7.87 \times 2 \times 0.2 + 1.8 \times 10^{-3} \times 2\pi \times 200 = 6.98 \text{ MeV}.$$

The initial muon energy is

$$E = \sqrt{p^2 + m^2} \simeq \sqrt{300^2 + 106^2} \text{ MeV} \simeq 318 \text{ MeV};$$

After one turn the energy becomes

$$E' = E - \Delta E \simeq 318 - 6.98 \text{ MeV} = 311 \text{ MeV}.$$

and the momentum is

$$p' = \sqrt{E'^2 - m^2} \simeq \sqrt{311^2 - 106^2} = 292 \text{ MeV}/c.$$

(a) The magnetic field needed to keep the muon in an orbit of radius R after one turn is

$$B' = \frac{p'[\text{GeV}/c]}{0.3 \, R[\text{m}]} \simeq 0.486 \text{ T}$$

and hence we have $\Delta B = B' - B \simeq 0.486 - 0.5 \simeq -0.014$ T.
(b) The muon mean decay pathlength is $\lambda = \beta\gamma c\tau$, where $\beta\gamma = p/m$. Hence we have

$$\lambda = \frac{p}{m} \, c\tau \simeq \frac{300 \times 3 \; 10^8 \times 2.2 \; 10^{-6}}{106} \simeq 1868 \text{ m}.$$

The mean number of turns is then

$$\langle n_{\text{turns}} \rangle = \frac{\lambda}{2\pi R} \simeq \frac{1868}{4\pi} \simeq 149.$$

Exercise 3.3.3

1. Muons come at rest in water loosing their kinetic energy by ionization. The energy lost by Cherenkov effect is negligible (order of per mill). Muons with 1 GeV/c momentum ($m = 106$ MeV/c^2) are close to the ionization minimum and we can use an energy loss rate of about 2 MeV/(g cm^{-2}) (= 2 MeV/cm in water). A simple estimate of the total pathlength can be done under the assumption that the changes in energy loss rate along the muon path can be neglected

$$R(E) \simeq \int_0^T \frac{dE}{(-dE/dl)_{\text{ion}}} \simeq \int_0^T \frac{dE}{2 \text{ MeV/cm}} \simeq \frac{900 \text{ MeV}}{2 \text{ MeV/cm}} \simeq 4.5 \text{ m},$$

where $T = \sqrt{p^2 + m^2} - m \simeq 900$ MeV is the initial kinetic energy of the muons.

A better estimate is made using the range-versus-energy plots reported in the PDG Review of Particle Physics [1]. Here only a few elements are shown: in particular for 1 GeV/c muons we get $R/m \approx 2000$ g cm^{-2} GeV^{-1} for H ($Z/A = 1$) and $R/m \approx 4000$ g cm^{-2} GeV^{-1} for C ($Z/A = 0.5$). In the mixtures, as water, one has to take into account that the primary dependence of the ionization energy loss is on the ratio Z/A. Therefore $-dE/dx$ of the mixture is proportional to $\langle Z/A \rangle = \sum w_j Z_j/A_j = \sum n_j Z_j / \sum n_j A_j$, where w_j (n_j) is the weight fraction (number of atoms) of the j-th element in the compound. In water we have $\langle Z/A \rangle = (2 \times 1 + 8)/(2 \times 1 + 16) \simeq 0.56$. Therefore the range in water is dominated by the energy loss in oxygen. If we take the range in carbon as a reference we obtain $R \simeq 4000$ cm/GeV $\times$ 0.106 GeV $\simeq 4.2$ m. We notice that the value obtained assuming a constant energy loss rate overestimates the actual range, but is adequate for a rough estimate.

2. The condition to emit Cherenkov photons is $\beta \geq \beta_{\min} = 1/n \simeq 0.75$. Hence we have for a particle mass m

$$p_{\min} = \beta_{\min} \gamma_{\min} m \implies p_{\min} \simeq 1.134\, m,$$

and then for a muon

$$T_{\min} = \sqrt{p_{\min}^2 + m^2} - m \simeq 0.51\, m \simeq 54 \text{ MeV}.$$

The length of the path where the muon emits Cherenkov radiation is (for a constant energy loss rate)

$$L_C = \int_{T_{\min}}^{T} \frac{dE}{(-dE/dl)_{\text{ion}}} \simeq \int_{T_{\min}}^{T} \frac{dE}{2 \text{ MeV/cm}} \simeq \frac{900 - 54 \text{ MeV}}{2 \text{ MeV/cm}} \simeq 4.23 \text{ m}.$$

Comparing this value with the one obtained under the same approximation we obtain a fraction $4.23/4.5 \simeq 94\%$.

3. The initial opening angle of the Cherenkov cone is obtained from

$$\cos \theta_C = \frac{1}{\beta n} = \frac{\sqrt{p^2 + m^2}}{p n} \simeq \frac{1.006}{1.33} \simeq 0.756.$$

The muon energy loss up to the exit from the detector is small enough (≈ 100 MeV) so that the Cherenkov angle is almost constant. Hence the region illuminated on the base is determined by the Cherenkov cone at the initial point and the radius of the circle is

$$R = D \tan \theta_C = \frac{D \sqrt{1 - \cos^2 \theta_C}}{\cos \theta_C} \simeq 0.865\, D \simeq 43 \text{ cm}.$$

Exercise 3.3.4

Considering the distance and the size of the detector, electrons and positrons are detected for angles between

$$\theta_{min} = \arctan(6/200) \simeq 30 \text{ mrad}$$

and

$$\theta_{max} = \arctan(10/200) \simeq 50 \text{ mrad}.$$

The energy of each beam is $E_e = \sqrt{s}/2 = 45 \text{ GeV}$.

Integrating the given expression of the Bhabha cross section between θ_{min} and θ_{max} we have

$$\sigma = \frac{8\pi\alpha^2}{E_e^2}(\hbar c)^2 \int_{\theta_{min}}^{\theta_{max}} \frac{d\theta}{\theta^3} = \frac{8\pi\alpha^2}{E_e^2}(\hbar c)^2 \left(\frac{1}{2 \cdot \theta_{min}^2} - \frac{1}{2 \cdot \theta_{max}^2} \right) =$$

$$\simeq 2.57 \cdot 10^{-8} \text{ fm}^2 \left(\frac{1}{2 \cdot (0.030)^2} - \frac{1}{2 \cdot (0.050)^2} \right) \simeq 0.91 \cdot 10^{-5} \text{ fm}^2 \simeq 9.1 \cdot 10^{-32} \text{ cm}^2$$

For a rate of 1 ev/s we obtain a luminosity

$$L = \frac{n}{\sigma} \simeq \frac{1 \text{ s}^{-1}}{9.1 \cdot 10^{-32} \text{cm}^2} \simeq 1.1 \cdot 10^{31} \text{ cm}^{-2}\text{s}^{-1}$$

Exercise 3.3.5

If τ is the mean lifetime, the number of particles surviving after a time t is

$$N(t) = N_0 e^{-t/\tau}.$$

In our case we require to have at least one decay in $t = 1 \text{ yr}$

$$N_0 - 1 = N_0 e^{-t/\tau} \simeq N_0 \left(1 - \frac{t}{\tau} \right)$$

hence $N_0 = \frac{\tau}{t}$. Since the number of nucleons in the detector mass M is $N_0 = N_A M$ (where $N_A \simeq 6.02 \cdot 10^{23} \text{ mole}^{-1}$ is the Avogadro number) we obtain for the required mass

$$M = \frac{\tau}{t \, N_A} \simeq \frac{10^{32}}{6.02 \cdot 10^{23} \text{ g}^{-1}} \simeq 1.7 \cdot 10^8 \text{ g} = 170 \text{ ton}$$

Exercise 3.3.6

The pion momentum in GeV/c is given by the relationship

$$p = 0.3 \, BR,$$

where R is the curvature radius in metres and B the magnetic field in Tesla. The deflection angle in the magnetic field can be written to a good approximation as

$$\theta \simeq \frac{L}{R} = \frac{0.3 \ BL}{p}.$$

We notice that the approximation of using L equal to the length of the magnet (instead of the length of the trajectory) is justified by the fact that the deflection angles are small for all the momenta (130 mrad at 0.5 GeV/c down to 44 at 1.5 GeV/c).

The distance L and the width w of the slit allow to select the pion momentum and its uncertainty. The angle subtended for a momentum $p_0 \pm \Delta p$ is

$$\Delta \theta \simeq \frac{w}{d} = \frac{0.3 \ BL}{p_0^2} \Delta p.$$

Hence to select 1 GeV/c $\pm 5\%$ charged pions we need a distance

$$d = \frac{p_0 \ w}{0.3 \ BL \left(\frac{\Delta p}{p_0} \right)} \simeq \frac{1 \times 0.01}{0.3 \times 0.2 \times 1.1 \times 0.1} \simeq 1.5 \text{ m}$$

Exercise 3.3.7

The relationship for a particle of charge e among the curvature radius R in metres, the uniform magnetic field B in Tesla and the momentum p in GeV/c is

$$p = 0.3 \ B \ R$$

hence the muon momentum is $p \simeq 2.1$ GeV/c.

For the revolution period we have

$$T = \frac{2\pi R}{\beta c},$$

where the velocity is $\beta = p/E = p/\sqrt{p^2 + m_\mu^2} \approx 1$. The revolution period is then $T \simeq 2.93 \times 10^{-7}$ s.

The mean muon lifetime in the Lab system is

$$\tau_{LS} = \gamma \tau \simeq \frac{p}{m_\mu} \tau \simeq 20 \ \tau.$$

The number of muons surviving after one period is

$$N(T) = N_0 e^{-T/\tau_{LS}},$$

where N_0 is the initial muon number. The fraction of muons decayed after one period is then

$$f = \frac{N_0(1 - e^{-T/\tau_{LS}})}{N_0} \simeq 1 - \left(1 - \frac{T}{\tau_{LS}}\right) = \frac{T}{\tau_{LS}} \simeq 6.7 \times 10^{-3}.$$

Exercise 3.3.8

1. The total energy of the particles involved in the proton decay is $E_0 = m_p$. In the considered decay channel the energies of the two particles are almost equal so that $\epsilon_e \approx \epsilon_{\pi^0} \approx E_0/2$ (the correct calculation gives 0.46 GeV and 0.48 GeV for the energies of the positron and pion respectively). Positrons and photons (from π^0 decay) are produced back-to-back and, having energies above the water critical energy and then produce e.m. cascades. Under the approximation of equal energies, the maximum of the longitudinal development is (in units of X_0)

$$T_{max} = \frac{\ln[E_0/(2E_c)]}{\ln 2} \simeq 2.55.$$

Hence most of the Cherenkov emitting particles are contained in a segment of length $L = 2 \times \frac{X_0}{\rho} \times T_{max} \simeq 1.8$ m. This length determines the size of the detector (each side $\gg L$).

2. To estimate the number of emitted Cherenkov photons, we need to evaluate the total track length for the charged particles (e^+, e^-) contained each cascade. This total length (called *track length integral*) is given by

$$T_{tot} = \frac{2}{3} \int_0^{T_{max}} 2^t \, dt = \frac{2}{3 \ln 2}(2^{T_{max}} - 1) = \frac{2}{3 \ln 2}\left(\frac{E_0}{2E_c} - 1\right) \simeq 4.7,$$

where the factor 2/3 is the average fraction of charged particles in the cascade. Hence we have for the total number of Cherenkov photons

$$N_{phot} = 2\frac{X_0 \cdot T_{tot}}{\rho} \times I_0 \simeq 338 \text{ cm} \times 400 \text{ cm}^{-1} \simeq 1.4 \times 10^5$$

Exercise 3.3.9

When a proton interacts with a residual air molecule it is thrown away from the trajectory where the accumulated protons are kept by the magnetic field. Then at each scattering a proton is lost. The absorption coefficient is given by

$$\mu = \sigma n \qquad n = \rho\frac{N_A}{A},$$

where σ is the total cross section, n is the number of scatterers per unit volume and N_A is the Avogadro number. In the proton ring (10^{-11} atm) we have

$$\mu = 300 \times 10^{-27} \times 1.25 \; 10^{-14} \times \frac{6 \; 10^{23}}{14} \simeq 1.6 \times 10^{-16} \; \text{cm}^{-1}.$$

The inverse of this value corresponds is the mean pathlength. 300 GeV protons are ultra-relativistic and their velocity is $\approx c$. Hence the mean beam lifetime is

$$\tau = \frac{1}{c\mu} \simeq 2.08 \times 10^5 \; \text{s} \simeq 58 \; \text{h}$$

Exercise 3.3.10

The interactions occurs against the nuclei along the beam. These are

$$N_{\text{sc}} = \frac{N_A}{A} \rho \, d \, S = \frac{6.02 \times 10^{23}}{207} \times 11.3 \times 0.2 \times \pi \times 1 \simeq 2 \times 10^{22}$$

The fraction of scattered particle is given by

$$f_s = \frac{N_s}{S} \sigma = \frac{2 \times 10^{22}}{3.14} \times 3 \; 10^{-26} \simeq 1.9 \; 10^{-4}$$

Exercise 3.3.11

The neutrino interaction rate is given by

$$w_{\text{int}} = \sigma\phi \simeq 7 \times 10^{-44} \; \text{cm}^2 \times 10^6 \text{cm}^{-2}\text{s}^{-1} \simeq 7 \times 10^{-38} \; \text{s}^{-1}.$$

The number of scattering centres (electrons) per unit volume is

$$n_{\text{sc}} = \rho \frac{Z}{A} N_A V = \frac{Z}{A} N_A M = 0.5 \times 6.02 \cdot 10^{23} \times 5 \cdot 10^{10} \simeq 1.5 \times 10^{34}.$$

Hence the number of interactions per year is ($\Delta T = 1 \; \text{yr} \simeq 3.15 \times 10^7 \; \text{s}$)

$$N_{\text{yr}} = w_{\text{int}} \times n_{\text{sc}} \times \Delta T = 3.3 \times 10^4$$

Exercise 3.3.12

(a) The maximum shower development is reached at a depth

$$T = \frac{\log_{10}(E_0/E_c)}{\log_{10} 2} = \frac{\log_{10}(500 \; \text{GeV}/80 \; \text{MeV})}{\log_{10} 2} \simeq 12.6$$

where T is expressed in radiation length units. Therefore the actual depth in g/cm^2 is

$$X_{\text{max}} = T \times X_0 \simeq 470 \; \text{g/cm}^2$$

corresponding to an optimal altitude (for vertical showers)

$$h = -h_0 \ln\left(\frac{X_{max}}{X_v(0)}\right) \simeq 5300 \text{ m.}$$

There are sites suitable for such observations, e.g. in the Andes or in Tibet.

(b) Electrons at the shower maximum have an energy equal to the critical energy $E_c^{water} \simeq E_c^{atm} = 80$ MeV. At this energy the Cherenkov condition is fulfilled

$$n\beta = \frac{n}{\sqrt{1 + (m/p)^2}} = \frac{1.33}{\sqrt{1 + (0.511/80)^2}} > 1.$$

Hence Cherenkov photons can be used to detect shower events.

(c) Shower particles (photons and electrons) at the critical energy have equal probability to loose energy by ionization and bremsstrahlung. Therefore they are not energetic enough to produce e.m. cascades. The component of the shower which are already electrons mostly loose energy by ionization. Instead those which are photons have still enough energy for pair production (threshold energy $\simeq 1$ MeV) and can generate electrons of both signs with lower energies. To make an estimate of the path done by electron loosing energy in water we can calculate the residual range of electrons at the critical energy

$$\Delta x = \int_0^{E_c^{water}} \frac{dE}{\left(\frac{-dE}{dx}\right)_{ion}} \simeq \frac{E_c^{water}}{2 \text{ MeV}/(\text{g cm}^{-2})} = 40 \text{ g/cm}^2$$

and then $\Delta l \simeq 40$ cm. Hence electrons loose their whole energies in the water tanks, apart those which exit the tank and loose only a part of their energy.

Exercise 3.3.13

1. The reaction threshold for the proton kinetic energy is

$$T_{th} = \frac{(2m_p + m_J)^2 - (2m_p)^2}{2m_p} = 2m_J + \frac{m_J^2}{2m_p} \simeq 11.3 \text{ GeV}$$

2. Denoting with M the total CMS energy for protons of 28 GeV energy against target protons (at rest), we have

$$M \simeq \sqrt{2m_p E_p} \simeq 7.3 \text{ GeV.}$$

The final state in the reaction (3.2) is a three body system. Then the maximum and minimum energy of the J/ψ particle in the CMS are given by

$$\min : E_J^* = m_J \simeq 3.1 \text{ GeV}$$

$$\max : E_J^* = \frac{M^2 + m_J^2 - (2m_p)^2}{2M} \simeq 4.07 \text{ GeV.}$$

To obtain the maximum and minimum values in the Lab system we make a Lorentz transformation with the following β and γ values

$$\beta = \frac{p_p}{E_p + m_p} \simeq 0.967, \qquad \gamma = \frac{E_p + m_p}{M} \simeq 3.96$$

To calculate the minimum and maximum J/ψ energies in the Lab frame we consider the following cases

min: $E_J = \gamma\,(E_J^*[\min] + \beta \cdot 0) \simeq 12.3$ GeV
max/min: $E_J = \gamma\,(E_J^*[\max] - \beta \cdot p_J^*\,[\max]) \simeq 6.0$ GeV
max/max: $E_J = \gamma\,(E_J^*[\max]] + \beta \cdot p_J^*[\max]) \simeq 26.2$ GeV

It follows that the minimum and maximum energies are 6 and 26.2 GeV respectively.
3. The minimum opening angle $\theta_{\min}$ of the e^+e^- pair is obtained from

$$\sin\left(\frac{\theta_{\min}}{2}\right) = \frac{\sqrt{m_J^2 - 2m_e^2}}{E_J} \simeq \frac{m_J}{E_J}$$

Therefore the minimum angle is obtained for the maximum J/ψ energy, 26.2 GeV, and turns out to be $\theta_{\min} \simeq 13.6°$.
4. Electrons are ultra-relativistic ($p \simeq E$): hence the e^+e^- invariant mass is

$$M_{ee}^2 \simeq 4p^+ p^- \sin^2\frac{\Delta\theta}{2}$$

where $p^+(p^-)$ is the $e^+(e^-)$ momentum and $\Delta\theta$ is the opening angle of the observed pair. Using for M_{ee} the J/ψ mass we obtain

$$p^+ = \frac{m_J^2}{4p^- \sin^2\frac{\Delta\theta}{2}} \simeq 12.4 \text{ GeV}/c.$$

Exercise 3.3.14

(a) The collider system is the CMS, hence $E_\tau = E_0/2 = 14.5$ GeV.
(b) Using the *Sargent rue* we have for the transition rates ($\Gamma = 1/\tau$)

$$\frac{\Gamma(\tau^+ \to e^+ + \nu_e + \bar{\nu}_\tau)}{\Gamma(\mu^+ \to e^+ + \nu_e + \bar{\nu}_\mu)} = \frac{m_\tau^5}{m_\mu^5}.$$

Taking into account the tau branching ratio into neutrinos we have

$$\Gamma(\tau^+ \to e^+ + \nu_e + \bar{\nu}_\tau) = \frac{BR(\tau^+ \to e^+ + \nu_e + \bar{\nu}_\tau)}{\tau_\tau}$$

Hence the tau mean lifetime is

$$\tau_\tau = \frac{BR(\tau^+ \to e^+ + \nu_e + \bar{\nu}_\tau)}{\Gamma(\tau^+ \to e^+ + \nu_e + \bar{\nu}_\tau)} = \tau_\mu \times BR(\tau^+ \to e^+ + \nu_e + \bar{\nu}_\tau) \times \left(\frac{m_\mu}{m_\tau}\right)^5 \simeq$$

$$\simeq 2.2 \ 10^{-6} \times 0.18 \times \left(\frac{106}{1777}\right)^5 \simeq 3.0 \times 10^{-13} \text{ s.}$$

(c) The τ mean pathlength is

$$\langle L \rangle = \beta \gamma c \tau_\tau = \frac{p_\tau}{m_\tau} c \tau_\tau = \frac{\sqrt{E_\tau^2 - m_\tau^2}}{m_\tau} c \tau_\tau \simeq 8.1 \times 3 \ 10^{10} \times 3 \ 10^{-13} \simeq 0.073 \text{ cm.}$$

For a cylindrical detector the minimum distance to observe a decay is given by the internal radius r (the distance increases with the angle). Hence the maximum detection probability is

$$f_{\max}(l > r) = \frac{1}{\langle L \rangle} \int_r^\infty e^{-\frac{l}{\langle L \rangle}} \, dl = e^{-\frac{r}{\langle L \rangle}} \simeq 2 \ 10^{-30}$$

and is then negligible.

Exercise 3.3.15

Photon and electron beams of 5 GeV produce electromagnetic showers. For their developments the relevant parameter is the number of radiation lengths. Each scintillator layer has $1/42 \simeq 0.02$ radiation lengths, whereas the lead slabs have about 2 radiation lengths each. Therefore the scintillator layers have a negligible contribution. Upstream of the fourth scintillator there are 3 lead slabs, hence the total number of radiation lengths is

$$T = \frac{3 \times 1 \text{ cm}}{0.56 \text{ cm}} \simeq 5.4.$$

Using the Heitler toy model the number of shower particles (e^+, e^- and γ) is 2^T. The scintillator detects charged particles via the ionization process whereas photons have a very low probability to convert to electrons because of the low Z of the material. In an e.m. showers charged particle are approximately 2/3 of the total content of particles, they have an energy loss rate corresponding to minimum ionizing particles ($\simeq 2$ MeV g^{-1} cm^2) and then their total energy release is

$$\Delta E = \frac{2}{3} \times 2^T \times \left(-\frac{dE}{dx}\right)_{\text{ion}} \times \rho d \simeq 0.67 \times 42.2 \times 2 \times 1.03 \simeq 58 \text{ MeV,}$$

where d is the scintillator thickness. This energy release is the same for incident electron and photons.

Instead muons loose energy only by ionization. The energy lost before the fourth scintillator is 2 MeV g^{-1} cm^2 $\times$ 3 cm $\times$ 11 g/cm^3 $\simeq 66$ MeV, hence their energy is almost unaffected. We can assume for them the same energy loss rate of $\simeq 2$ MeV g^{-1} cm^2 and then the energy release in the fourth scintillator is

$$\Delta E = 2 \times 1.03 \simeq 2 \text{ MeV.}$$

Finally to discriminate electrons from photons we can use the signal in the first scintillator which can be detected only for electrons but is absent for photons.

Exercise 3.3.16

(a) The p-Cu interaction length is

$$\lambda_{int}^{pCu} = \frac{A}{N_A \, \rho \, \sigma_{pCu}} = \frac{A^{\frac{1}{3}}}{N_A \, \rho \, \sigma_{pp}} = \frac{63.5^{\frac{1}{3}}}{6 \; 10^{23} \; 8.96 \; 40 \; 10^{-27}} \simeq 18.5 \text{ cm}$$

(b) The initial state has baryon number $B = +2$, the two D-particles are mesons and have $B = 0$. Hence X must have $B = +2$. The simplest case is

$$p + p \to D^+ + D^- + p + p$$

The flavor flux diagram is shown in Fig. 3.2 (left).

(c) The quark flavor content of D^+ and D^- is $D^+ = c\bar{d}$ and $D^- = \bar{c}d$ respectively. D^+ decays into neutrinos via $c \to W^+ + s$ followed by $W^+ \to l^+ + \nu_l$. Hence D^+ is associated to neutrinos. Similarly D^- decays into W^- and then anti-neutrinos are produced. Examples of Feynman diagrams with $\bar{\nu}_e$ and ν_μ final states are shown in Fig. 3.2 (right).

(d) The interaction length of D-particles is

$$\lambda_{DCu} = \frac{\sigma_{pp}}{\sigma_{Dp}} \times \lambda_{pCu} = \frac{40}{30} \times 18.5 \text{ cm} \simeq 25 \text{ cm}$$

The decay length is instead

$$\lambda_{dec} = \beta\gamma \; c\tau(D^\pm) = \frac{p}{m_{D^\pm}} \times c\tau(D^\pm)$$

Therefore $\lambda_{dec} \ll \lambda_{DCu}$ is obtained for

$$p \ll m_{D^\pm} \times \frac{\lambda_{DCu}}{c\tau(D^\pm)} \simeq 1.87 \times \frac{25}{3 \; 10^{10} \times 1.04 \; 10^{-12}} \text{ GeV/c} \simeq 1500 \text{ GeV/c}$$

which is always fulfilled for 400 GeV incident protons.

(e) Taking into account the considerations at point (c) and the fact that $BR(D^\pm \to \nu_\mu/\bar{\nu}_\mu) = BR(D^\pm \to \nu_e/\bar{\nu}_e)$ we expect for the muon to electron neutrino ratio

$$\frac{\nu_\mu + \bar{\nu}_\mu}{\nu_e + \bar{\nu}_e} = 1.$$

Exercise 3.3.17

The absorption coefficient for pair production, which is the dominant process at high energies, is given by

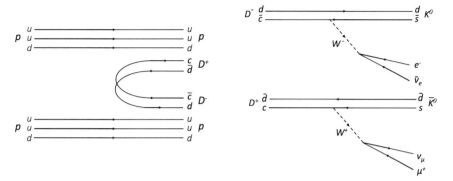

Fig. 3.2 Flavor flux diagram for $p + p \rightarrow D^{+} + D^{-} + p + p$ (left). Feynman diagrams for two D decays (right)

$$\mu = \left(\frac{7}{9}\right) X_0^{-1} \simeq 1.38 \text{ cm}^{-1}$$

A photon hitting the lead plate has a probability $\exp(-\mu d)$ to escape from the lead plate. Instead in case of pair production electrons emerge from the plate or induce e.m. showers, depending on the first interaction point. If conversion occurs one or more electrons will reach the downstream detector. The conversion probability is then

$$P_c = 1 - \exp(-\mu d)$$

Hence the π^0 detection efficiency is

$$\epsilon = P_c^2 = [1 - \exp(-\mu d)]^2 = [1 - \exp(-1.38)]^2 \simeq 56\%$$

Exercise 3.3.18

(a) The process that makes electron antineutrino detectable is the same used in the celebrated experiment by Reines and Cowan [2]

$$\bar{\nu}_e + p \rightarrow n + e^+$$

The process, called also 'inverse beta decay', is a charged current weak interaction (i.e. with W virtual boson). The detected particles are the positron through its annihilation in two photons of 0.5 MeV and the delayed photons emitted by the capture of the neutron. The process has the cross section given in the text. Instead muon antineutrinos, originating from the oscillation phenomenon, are difficult to detect. In fact the charged current process $\bar{\nu}_\mu + p \rightarrow n + \mu^+$ is forbidden by kinematics ($E_{\text{thr}} \simeq 100$ MeV) and the neutral current process (i.e. with Z^0 virtual boson) $\bar{\nu}_\mu + p/n \rightarrow \bar{\nu}_\mu + p/n$ can be only detected from the nucleon recoil with very low efficiency. Hence the oscillation phenomenon can be observed counting the number of disappeared electron antineutrinos.

(b) If neutrinos do not oscillate the interaction rate is

$$r = \frac{I_\nu}{4\pi L^2} \times \sigma_\nu \times N_n \times \rho l S$$

where N_n is the number of target nucleons per gram, ρ is the medium density, l and S are the detector length and section respectively. The product $\rho l S$ is the detector mass and we have

$$r = \frac{I_\nu}{4\pi L^2} \times \sigma_\nu \times N_A \times M \simeq$$

$$\simeq \frac{10^{18}\ \mathrm{s}^{-1}}{12.56\ 20000^2\ \mathrm{cm}^2} \times 2\ 10^{-43}\ \mathrm{cm}^2 \times \frac{6\ 10^{23}}{\mathrm{g}} \times 10^6\ \mathrm{g} \simeq 2.4\ 10^{-5}\ \mathrm{s}^{-1}$$

Denoting with ϵ the detection efficiency, the number of expected interactions per year is

$$N = r \times \epsilon \times T \simeq 2.4\ 10^{-5} \times 0.70 \times 3.15\ 10^7 \simeq 529.$$

(c) For a detector at 200 m from the reactor core and 2 MeV electron antineutrinos the probability to become muon antineutrinos is

$$P(\bar{\nu}_e \to \bar{\nu}_\mu) \simeq 0.20\ \sin^2\left(10^{-3}\ \frac{L[\mathrm{m}]}{E[\mathrm{MeV}]}\right) \simeq 0.20 \times \sin^2\left(10^{-3}\ \frac{200}{2}\right) \simeq 0.002$$

Hence the number of detectable electron antineutrinos is

$$N_e \simeq N \times [1 - P(\bar{\nu}_e \to \bar{\nu}_\mu)] \simeq 528$$

and the mean number of disappeared $\bar{\nu}_e$ is 1.1.

(d) The probability to have a null result is given by the poissonian probability to observe no event out of an expectation of 1.1

$$P(0|2) = e^{-1.1}\frac{1.1^0}{0!} = e^{-1.1} \simeq 33\%$$

It is worth to notice that this probability is not realistic, because it is based on the assumption that the knowledge of the number of neutrino interactions is perfectly known. In real experiments the uncertainty on the neutrino flux and detection efficiency makes it impossible to observe a disappearance ratio (1/529) so small.

Exercise 3.3.19

(a) To get the mass of the particle we consider the region 2, after the slowing down, where two measurements are available.

From the time-of-flight we have $\beta_2 = \frac{v_2}{c} \simeq \frac{2.8 \times 10^8}{3 \times 10^8} \simeq 0.93$ and $\beta_2 \gamma_2 = \frac{\beta_2}{\sqrt{1 - \beta_2^2}} \simeq$ 2.60.

From the curvature we have $p_2 = 0.3\, B R_2 = 0.3 \times 1 \times 1.21 \simeq 0.363 \text{ GeV/c}$. The rest mass of the particle is

$$m = \frac{p_2}{\beta_2 \gamma_2} \simeq \frac{0.363}{2.60} \simeq 0.140 \text{ GeV/c}^2.$$

It is a charged pion whose momentum before the slowing down is

$$p_1 = 0.3\, B \frac{l_1^2}{8 s_1} \simeq 0.3 \times 1 \times \frac{0.80^2}{8 \times 0.03} \simeq 0.80 \text{ GeV/c},$$

and the kinetic energy is

$$T_1 = \sqrt{p_1^2 + m^2} - m \simeq 0.670 \text{ GeV}.$$

(b) The energy lost in the medium is

$$\Delta E = T_1 - T_2 = T_1 - \left(\sqrt{p_2^2 + m^2} - m \right) \simeq 0.670 - 0.250 \simeq 0.420 \text{ GeV}.$$

(c) The mean half-time is $T_{1/2} = L_{1/2}/(c\beta_2\gamma_2)$. Then the mean lifetime is

$$\tau = \frac{L_{1/2}}{c\beta_2\gamma_2 \ln 2} \simeq \frac{14}{3 \times 10^8 \times 2.6 \times 0.69} \simeq 2.6 \times 10^{-8} \text{ s}.$$

Exercise 3.3.20

(a) Neglecting energy losses, the momentum of the electron (positron) is

$$\frac{p}{\text{GeV/c}} = 0.3 \frac{B}{\text{Tesla}} \frac{R}{\text{m}} \qquad (3.10)$$

$$\simeq 0.3 \times 0.8 \times 0.40 \simeq 0.096.$$

The photon energy is the sum of the two momenta $E_\gamma = 2\, p \simeq 192 \text{ MeV}$. In this calculation the opening angle of the pair has not been considered: in fact it is negligible, $\theta \approx m_e/E_\gamma \simeq 2.7 \times 10^{-3}$.

(b) To make a rough estimate of the energy loss along the electron (positron) track, we assume that the track length is the same as in the previous case (though the track actually changes). This energy loss is due to ionization, because the bremsstrahlung is negligible for $E < E_{\text{crit}}$ ($\approx 300 \text{ MeV}$ in LH_2)

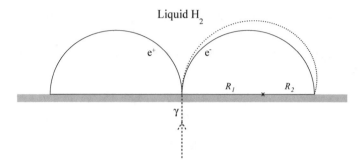

Fig. 3.3 Solid line: no energy losses; dashed line: with energy losses

$$\Delta E \simeq \left(-\frac{dE}{dx} \right)_{\text{ion}} \rho \, \pi \, R$$

Since the electron (positron) has p/m_e of few hundreds, we can assume[9] that $(-dE/dx)_{\text{ion}} \approx (-dE/dx)_{\text{min}} \simeq 4.1$ MeV g^{-1} cm^2. Therefore we have

$$\Delta E \simeq 4.1 \times 0.071 \times 125.6 \simeq 36.6 \text{ MeV}.$$

The electron (positron) momentum at the entrance of the chamber can be estimated using the following arguments (see Fig. 3.3):

- the track is not a semi-circle: its radius at entrance R_1 (at exit R_2) is larger (smaller) than the circle radius in the case of no energy losses R;
- the sum of these two radii can be approximated to $2R$ (the measured diameter).

Since $R \propto p$, denoting with p_{in} and p_{out} respectively the momentum at entrance and exit of the chamber, from $2R = R_1 + R_2$ we obtain

$$p = \frac{p_{\text{in}} + p_{\text{out}}}{2} = \frac{2p_{\text{in}} - \Delta p}{2}.$$

Therefore we have:

$$p_{\text{in}} \simeq p + \frac{\Delta p}{2} \simeq p + \frac{\Delta E}{2} \simeq 96 + \frac{37}{2} \simeq 114.5 \text{ MeV}$$

and the photon energy is $E_\gamma^{\text{corr}} = 2 \, p_{\text{in}} \simeq 229$ MeV.

A more accurate calculation can be done as follows.

$$dp \simeq \left(-\frac{dE}{dx} \right)_{\text{ion}} \rho dl \simeq \left(-\frac{dE}{dx} \right)_{\text{ion}} \rho \, R d\alpha.$$

---

[9]This value can be obtained from either the table "Atomic and nuclear properties of materials" in the PDG Review of Particle Physics [1] or simply from a standard value (for light materials) of 2 MeV g^{-1} cm^2 multiplied by 2 because of the Z/A ratio for the Hydrogen.

In this expression we have assumed that the arc element is centered as in the case of no losses: it is not actually true, but it is a sensible approximation for an estimate. Substituting here Eq. (3.10) one gets

$$\frac{dp}{p} = \left(-\frac{dE}{dx}\right)_{\text{ion}} \rho \frac{3}{B} \, d\alpha = -k \, d\alpha,$$

with $k = 4.1 \times 0.071 \times 3/0.8 \simeq 0.121$. Integrating we have

$$p(\alpha) = p_{\text{in}} \exp(-k\alpha).$$

Therefore

$$\Delta p = p_{\text{in}} - p_{\text{out}} = p_{\text{in}} [1 - \exp(-k\pi)]$$

$$p_{\text{in}} = \frac{\Delta p}{1 - \exp(-k\pi)} \simeq \frac{37}{1 - \exp(-0.121\pi)} \simeq 117 \text{ MeV}$$

from which we get $E_\gamma^{\text{corr}} = 2 \, p_{\text{in}} \simeq 234$ MeV.

Exercise 3.3.21

(a) The dominating process at this energy is Compton scattering by which photons transfer part of their energies to electrons. Iterating this process the whole energy of the photons is deposited and the measurement is possible through the ionization energy loss of the electrons. The characteristic length which is relevant to determine the sizes of the detector is the Compton mean free path

$$\lambda_C = \frac{A}{Z} \frac{1}{N_A \, \sigma_C},$$

where σ_C is the Compton cross section, N_A is the Avogadro number and A, Z refer to the detector material. To make a rough estimate one can assume $A/Z \approx 2$ and use the Thomson cross section, σ_T, for the Compton scattering

$$\lambda_C \approx 2 \times \frac{1}{6 \times 10^{23} \times 6.6 \times 10^{-25}} \simeq 5 \text{ g cm}^{-2}.$$

A more accurate calculation would give a larger λ_C (by about a factor 2), being the Thomson cross section the low energy limit of Compton scattering.

(b) In the antineutrino scattering $\bar{\nu}_e + p \rightarrow e^+ + n$, the outgoing particles have momenta of the same order of the momentum of the incident neutrino. Assuming that $p_n \approx E_\nu$, the neutron is non-relativistic and we get for the its kinetic energy

$$T_n \approx \frac{E_\nu^2}{2 \, m_n} \simeq 2 \text{ keV}$$

which means that the recoil energy is negligible with respect to the other energies.[10]
(c) Denoting by E_+ the positron energy, from energy conservation in the process
$\bar{\nu}_e + p \rightarrow e^+ + n$, neglecting the neutron recoil energy, we get

$$E_\nu = E_+ + m_n - m_p.$$

Energy conservation applied to the positron annihilation gives

$$E_{\text{vis}} = E_+ + m_e. \tag{3.11}$$

Then the asked relationship is

$$E_\nu = E_{\text{vis}} + m_n - m_p - m_e \simeq E_{\text{vis}} + 0.78 \text{ MeV}$$

(d) Since $E_{\text{vis}} \geq 2m_e$ because of Eq. (3.11), the detected neutrinos must have

$$E_\nu \geq 2m_e + 0.78 \text{ MeV} \simeq 1.78 \text{ MeV}.$$

This corresponds to the energy threshold of the process.

References

1. Tanabashi, M., et al.: (Particle data group). Phys. Rev. D **98**, 030001 (2018). http://pdg.lbl.gov/
2. Reines, F., Cowan, Jr., C.L.: Free anti neutrino absorption cross section. I. Measurement of the free anti neutrino absorption cross section by protons. Phys. Rev. **113**, 273 (1959)

---

[10]The accurate calculation can be done in a specific case, e.g. the maximum kinetic energy: it is obtained for a neutron emitted in the forward direction. In this case we have all momenta along the same direction. Note that the calculation requires to use accurate mass values for the three particles: $m_p = 938.272$, $m_n = 939.565$, $m_e = 0.511$, all in MeV/c^2. The invariant mass squared is $M^2 = m_p^2 + 2\, m_p\, E_\nu$ and then $M \simeq 940.27$ MeV. The neutron energy in the CMS is

$$E_n^* = \frac{M^2 + m_n^2}{2\, M} \simeq 939.565 \text{ MeV},$$

and the corresponding momentum is $p^* \simeq 0.48$ MeV/c. The β of the CM is $\beta = E_\nu/(E_\nu + m_p) \simeq 0.021$ and then the Lorentz factor is 1. The neutron momentum in the Lab is

$$p_n = \gamma(p^* + \beta E_n^*) \simeq 0.48 + 0.0021 \times 939.565 \simeq 2.5 \text{ MeV/c},$$

which corresponds to a neutron kinetic energy 3.2 keV.

Printed in the United States
By Bookmasters